The Book of Nothing

By the Same Author

Theories of Everything

The Left Hand of Creation
(with Joseph Silk)

The Anthropic Cosmological Principle
(with Frank J. Tipler)

The World Within the World

The Artful Universe

Pi in the Sky

Impossibility

The Origin of the Universe

Between Inner Space and Outer Space

The Universe that Discovered Itself

The Book
of Nothing

Vacuums, Voids, and the Latest Ideas
about the Origins of the Universe

John D. Barrow

Pantheon Books, New York

Library of Congress Cataloging-in-Publication Data

Barrow, John D., 1952–
The book of nothing : vacuums, voids, and the latest ideas about the origins
of the universe / John D. Barrow.
p.   cm.
Includes index.
ISBN 0-375-42099-1
1. Zero (The Number)   2. Vacuum.   3. Nothing (Philosophy)   I. Title.
QA141.B36   2001   111'.5—dc21   00-058894

www.pantheonbooks.com

Printed in the United States of America

First American Edition

In Memory of Dennis Sciama

"From time to time young men have brought me, for my advice, pieces of work, pretty well completed but with no title. This amazes me, for to me the title is the compass-setting by which the whole development is steered. Of course, I may change it on second thoughts, but then whatever emerges in the end is schizophrenic — one cannot serve two masters."

John L. Synge
Talking About Relativity

Contents

Preface

"Deciding on a book's beginning is as complex as determining the origins of the universe."

Robert McCrum

'Because it's not there' might be reason enough to write a book about
Nothing, especially if the author has already written one about Everything.
But, fortunately, there are better reasons than that. If one looks at the special problems that were the mainsprings of progress along the oldest and
most persistent lines of human inquiry, then one finds Nothing, suitably
disguised as something, never far from the centre of things.

Nothing, in its various guises, has been a subject of enduring fascination for millennia. Philosophers struggled to grasp it, while mystics
dreamed they could imagine it; scientists strove to create it; astronomers
searched in vain to locate it; logicians were repelled by it, yet theologians
yearned to conjure everything from it; and mathematicians succeeded.
Meanwhile, writers and jesters were happy to stir up as much ado about
Nothing as ever they possibly could. Along all these pathways to the truth
Nothing has emerged as an unexpectedly pivotal something, upon which so
many of our central questions are delicately poised.

Here, we are going to draw together some of the ways in which our
conceptions of Nothing influenced the growth of knowledge. We will see
how the ancient Western addiction to logic and analytic philosophy prevented progress towards a fruitful picture of Nothing as something that
could be part of an explanation for the things that are seen. By contrast,
Eastern philosophies provided habits of thought in which the idea of
Nothing-as-something was simple to grasp and not only negative in its

ramifications. From this first simple step, there followed a giant leap for mankind: the development of universal counting systems that could evolve onwards and upwards to the esoteric realms of modern mathematics.

In science, we will see something of the quest to make a real vacuum, in the midst of a thousand years of tortuous argument about its possibility, desirability and place. These ideas shaped the future direction of many parts of physics and engineering while, at the same time, realigning the philosophical and theological debates about the possibility and desirability of the vacuum – the physical Nothing. For the theologians, these debates were, in part, the continuation of a crucial argument about the need for the Universe to have been created out of both a physical and a spiritual Nothing. But for the critical philosophers, they were merely particular examples of ill-posed questions about the ultimate nature of things that were gradually falling into disrepute.

At first, such questions about the meaning of Nothing seemed hard, then they appeared unanswerable, and then they appeared meaningless: questions about Nothing weren't questions about anything. Yet, for the scientists, producing a vacuum appeared to be a physical possibility. You could experiment with the vacuum and use it to make machines: an acid test of its reality. Soon this vacuum seemed unacceptable. A picture emerged of a Universe filled with a ubiquitous ethereal fluid. There was no empty space. Everything moved through it; everything felt it. It was the sea in which all things swam, ensuring that no nook or cranny of the Universe could ever be empty.

This spooky ether was persistent. It took an Einstein to remove it from the Universe. But what remained when everything that could be removed was removed was more than he expected. The combined insights of relativity and the quantum have opened up striking new possibilities that have presented us with the greatest unsolved problems of modern astronomy. Gradually, over the last twenty years, the vacuum has turned out to be more unusual, more fluid, less empty, and less intangible than even Einstein could have imagined. Its presence is felt on the very smallest and largest dimensions over which the forces of Nature act. Only when the vacuum's subtle quantum influence was discovered could we see how the diverse forces of Nature might unite in the seething microworld inhabited by the most elementary parts of matter.

The astronomical world is no less subservient to the properties of the vacuum. Modern cosmology has built its central picture of the Universe's past, present and future on the vacuum's extraordinary properties. Only time will tell whether this construction is built on shifting sand. But we may not have to wait very long. A series of remarkable astronomical observations now seem to be revealing the cosmic vacuum by its effects on the expansion of the Universe. We look to other experiments to tell us whether, as we suspect, the vacuum performed some energetic gymnastics nearly fifteen billion years ago, setting the Universe upon the special course that led it to be what it is today and what it will eventually become.

I hope that this story will convince you that there is a good deal more to Nothing than meets the eye. A right conception of its nature, its properties, and its propensity to change, both suddenly and slowly, is essential if we are to understand how we got to be here and came to think as we do.

The glyphs accompanying the chapter numbers throughout this book, from zero to nine, are reproductions of the beautiful Mayan head-variant numerals. They represent a spectrum of celebrated gods and goddesses and were widely used by the Mayans more than fifteen hundred years ago for recording dates and spans of time.

I would like to thank Rachel Bean, Malcolm Boshier, Mariusz Dąbrowski, Owen Gingerich, Jörg Hensgen, Martin Hillman, Ed Hinds, Subhash Kak, Andrei Linde, Robert Logan, João Magueijo, Martin Rees, Paul Samet, Paul Shellard, Will Sulkin, Max Tegmark and Alex Vilenkin for their help and discussions at various times. This book is dedicated to the memory of Dennis Sciama without whose early guidance neither this, nor any of my other writing over the last twenty-five years, would have been possible.

This book has survived one move of house and three moves of office in the course of its writing. In the face of all these changes of vacuum state, I would also like to thank my wife Elizabeth for ensuring that something invariably prevailed over nothing, and our children, David, Roger and Louise, for their unfailing scepticism about the whole project.

J.D.B., Cambridge, May 2000

The Book of Nothing

Nothingology – Flying to Nowhere

"As I was going up the stair,
I met a man who wasn't there.
He wasn't there again today,
I wish, I wish he'd stay away."

Hughes Mearns

MYSTERIES OF NON-EXISTENCE

"You ain't seen nothing yet."

Al Jolson[1]

'Nothing', it has been said, 'is an awe-inspiring yet essentially undigested concept, highly esteemed by writers of a mystical or existentialist tendency, but by most others regarded with anxiety, nausea, and panic.'[2] Nobody seems to know how to handle it and perplexingly diverse conceptions of it exist in different subjects.[3] Just take a look at the entry for 'nothing' in any good dictionary and you will find a host of perplexing synonyms: nil, none, nowt,[4] nulliform,[5] nullity – there is a nothing for every occasion. There are noughts of all sorts to zero-in on, from zero points to zero hours, ciphers to nulliverses.[6] There are concepts that are vacuous, places that are evacuated, and voids of all shapes and sizes. On the more human side, there are nihilists,[7] nihilianists,[8] nihilarians,[9] nihilagents,[10] nothingarians,[11] nullifideans,[12] nullibists,[13] nonentities and nobodies. Every walk of life seems to have its own personification of nothing. Even the financial pages of my newspaper tell me that 'zeros'[14] are an increasingly attractive source of income.

Some zeros seem positively obscure, almost circumlocutory. Tennis can't bring itself to use so blunt a thing as the word 'nil' or 'nothing' or 'zero' to record no score. Instead, it retains the antique term 'love', which has reached us rather unromantically from *l'oeuf*, the French for an egg which represented the round 0 shape of the zero symbol.[15] Likewise, we still find the use of the term 'love' meaning 'nothing' as when saying you are playing for love (rather than money), hence the distinction of being a true 'amateur', or the statement that one would not do something 'for love or money', by which we mean that we could not do it under any circumstances. Other games have evolved anglicised versions of this anyone-for-tennis pseudonym for zero: 'goose egg' is used by American ten-pin bowlers to signal a frame with no pin knocked down. In England there is a clear tradition for different sports to stick with their own measure of no score, 'nil' in soccer, 'nought' in cricket, but 'ow' in athletics timings, just like a telephone number, or even James Bond's serial number. But sit down at your typewriter and 0 isn't O any more.

'Zilch' became a common expression for zero during the Second World War and infiltrated 'English' English by the channel of US military personnel stationed in Britain. Its original slang application was to anyone whose name was not known. Another similar alliterative alternative was 'zip'. A popular comic strip portrays an owl lecturing to an alligator and an infant rabbit on a new type of mathematics, called 'Aftermath', in which zero is the only number permitted; all problems have the same solution – zero – and consequently the discipline consists of discovering new problems with that inevitable answer.[16]

Another curiosity of language is the use of the term 'cipher' to describe someone who is a nonentity ('a cipher in his own household', as an ineffectual husband and father was once described). Although a cipher is now used to describe a code or encryption involving symbols, it was originally the zero symbol of arithmetic. Here is an amusing puzzle which plays on the double meaning of cipher as a code and a zero:

"U 0 a 0, but I 0 thee
O 0 no 0, but O 0 me.
O let not my 0 a mere 0 go,
But 0 my 0 I 0 thee so."

which deciphers to read

"You sigh for a cipher, but I sigh for thee
O sigh for no cipher, but O sigh for me.
O let not my sigh for a mere cipher go,
But sigh for my sigh, for I sigh for thee so."

The source of the insulting usage of cipher is simple: the zero symbol of arithmetic is one which has no effect when added or subtracted to anything. One Americanisation of this is characteristically racier and derives from modern technical jargon. A null operation is technospeak for an action that has no consequence. Your computer cycles through millions of them while it sits waiting for you to make the next keystroke. It is a neutral internal computer operation that performs no calculation or data manipulation. Correspondingly, to say that someone 'is a zero, a real null op' needs no further elucidation. Of course, with the coming of negative numbers new jokes are possible, like that of the individual whose personality was so negative that when he walked into a party, the guests would look around and ask each other 'who left?' or the scientist whose return to the country was said to have added to the brain drain. The adjective 'napoo', meaning finished or empty, is a contraction of the French *il n'y a plus*, for 'there is nothing left'.

Not all nominal associations with 'nothing' were derogatory. Sometimes they had a special purpose. When some of the French Huguenots fled to Scotland to escape persecution by Louis XIV they sought to keep their names secret by using the surname Nimmo, derived from the Latin *ne mot*, meaning no one or no name.

Our system of writing numbers enables us to build up expressions for numbers of unlimited size simply by adding more and more noughts to the right-hand end of any number: I I230000000000 ... During the hyperin-flationary period of the early 1920s, the German currency collapsed in value so that hundreds of billions of marks were needed to stamp a letter. The economist John K. Galbraith writes[17] of the psychological shock induced by these huge numbers with their strings of zeros:

> "'Zero stroke' or 'cipher stroke' is the name created by German physicians for a prevalent nervous malady brought about by the present fantastic currency figures. Scores of cases of the 'stroke' are reported among men and women of all classes, who have been prostrated by their efforts to figure in thousands of millions. Many of these persons apparently are normal, except for a desire to write endless rows of ciphers."

Pockets of hyperinflation persist around the globe; indeed there are more zeros around today than at any other time in history. The introduction of binary arithmetic for computer calculation, together with the profusion of computer codes for the control of just about everything, has filled machines with 0s and 1s. Once you had a ten per cent chance of happening upon a zero, now it's evens. But there are huge numbers that are now almost commonplace. Everyone knows there are billions and billions of stars, and national debts conjure up similar astronomical numbers. Yet we have found a way to hide the zeros: 10^9 doesn't look as bad as 1,000,000,000.

The sheer number of synonyms for 'nothing' is in itself evidence of the subtlety of the idea that the words try to capture. Greek, Judaeo-Christian, Indian and Oriental traditions all confronted the idea in different ways which produced different historical threads. We will find that the concept of nothingness that developed in each arena merely to fill some sort of gap then took on a life of its own and found itself describing a something that had great importance. The most topical example is the physicists' concept of nothing – the vacuum. It began as empty space – the void, survived Augustine's dilution to 'almost nothing',[18] turned into a stagnant ether through which all the motions in the Universe swam, vanished in Einstein's hands, then re-emerged in the twentieth-century quantum picture of how Nature works. This perspective has revealed that the vacuum is a complex structure that can change its character in sudden or gradual ways. Those changes can have cosmic effects and may well have been responsible for endowing the Universe with many of its characteristic features. They may have made life a possibility in the Universe and one day they may bring it to an end.

When we read of the difficulties that the ancients had in coming to terms with the concept of nothing, or the numeral for zero, it is difficult to put oneself in their shoes. The idea now seems commonplace. But mathematicians and philosophers had to undergo an extraordinary feat of mental gymnastics to accommodate this everyday notion. Artists took rather longer to explore the concepts of Nothing that emerged. But, in modern times, it is the artist who continues to explore the paradoxes of Nothing in ways that are calculated to shock, surprise or amuse.

NOTHING VENTURED

"Now, is art about drawing or is it about colouring in?"

Ali G

"Nothing is closer to the supreme commonplace of our commonplace age than its preoccupation with Nothing . . . Actually, Nothing lends itself very poorly indeed to fantastic adornment."

Robert M. Adams[19]

In the 1950s artists began to explore the limiting process of going from polychrome to monochrome to nullichrome. The American abstract artist

Ad Reinhardt produced canvases coloured entirely red or blue, before graduating to a series of five-foot-square all-black productions that toured the leading galleries in America, London and Paris in 1963. Not surprisingly, some critics condemned him as a charlatan[20] but others admired his art noir: 'an ultimate statement of esthetic purity', according to American art commentator Hilton Kramer.[21] Reinhardt went on to run separate exhibitions of his all-red, all-blue and all-black canvases and writes extensively about the *raison d'être* for his work.[22] It is a challenge to purists to decide whether Reinhardt's all-black canvases capture the representation of Nothing more completely than the all-white canvases of Robert Rauschenberg. Personally, I prefer the spectacular splash of colours in Jasper Johns' *The Number Zero*.[23]

The visual zero did not need to be explicitly represented by paint or obliquely signalled by its absence. The artists of the Renaissance discovered the visual zero for themselves in the fifteenth century and it became the centrepiece of a new representation of the world that allowed an infinite number of manifestations. The 'vanishing point' is a device to create a realistic picture of a three-dimensional scene on a flat surface. The painter fools the eye of the viewer by imagining lines which connect the objects being represented to the viewer's eye. The canvas is just a screen that intervenes between the real scene and the eye. Where the imaginary lines intersect that screen, the artist places his marks. Lines running parallel to the screen are represented by parallel lines which recede to the line of the distant horizon, but those seen as perpendicular to the screen are represented by a cone of lines that converge towards a single point – the vanishing point – which creates the perspective of the spectator.

Musicians have also followed the piper down the road to nothingtown. John Cage's musical composition *4′ 33″* – enthusiastically encored in some halls – consists of 4 minutes and 33 seconds of unbroken silence, rendered by a skilled pianist wearing evening dress and seated motionless on the piano stool in front of an operational Steinway. Cage explains that his idea is to create the musical analogue of absolute zero of temperature[24] where all thermal motion stops. A nice idea, but would you pay anything other than nothing to see it? Martin Gardner tells us that 'I have not heard *4′ 33″* per-formed but friends who have tell me that it is Cage's finest composition'.[25]

Writers have embraced the theme with equal enthusiasm. Elbert Hubbard's elegantly bound *Essay on Silence* contains only blank pages, as does a chapter in the autobiography of the English footballer Len Shackleton which bears the title 'What the average director knows about football'. An empty volume, entitled *The Nothing Book*, was published in 1974 and appeared in several editions and even withstood a breach of copyright action by the author of another book of blank pages.

Another style of writing uses Nothing as a fulcrum around which to spin opposites that cancel. Gogol's *Dead Souls* begins with a description of a gentleman with no characteristics arriving at a town known only as N.:

> "The gentleman in their carriage was not handsome but neither was he particularly bad-looking; he was neither too fat nor too thin; he could not be said to be too old, but he was not too young either."

A classic example of this adversarial descriptive style, in which attributes and counter-attributes cancel out to zero, is to be found on a woman's tomb in Northumberland. The family inscribed the words

> "She was temperate, chaste, and charitable, but she was proud, peevish, and passionate. She was an affectionate wife and tender mother but her husband and child seldom saw her countenance without a disgusting frown . . ."[26]

Not to be forgotten, of course, are those commercial geniuses who are able to make more out of nothing than most of us can earn from anything. 'Polo, the mint with the hole' is one of the best-known British advertising pitches for a sweet that evolved independently as a 'Lifesaver' in the United States. More than forty years of successful marketing have promoted the hole in the mint rather than the mint itself. Nobody seems to notice that they are buying a toroidal confection that contains a good chunk of empty space, but then he wouldn't.

NOTHING GAINED

"Nothing is real."

The Beatles, "Strawberry Fields Forever"

So much for these snippets of nothing. They show us nothing more than that there is a considerable depth and breadth to the contemplation of Nothing. In the chapters to come, we shall explore some of these unexpected paths. We shall see that, far from being a quirky sideshow, Nothing is never far from the central plots in the history of ideas. In every field we shall explore, we shall find that there is a central issue which involves a right conception of Nothing, and an appropriate representation of it. Philosophical overviews of key ideas in the history of human thought have always made much of concepts like infinity,[27] but little of Nothing. Theology was greatly entwined with the complexities of Nothing, to decide whether we were created out of it and whether we risked heading back into its Godless oblivion. Religious practices could readily make contact with the reality of Nothingness through death. Death as personal annihilation is an ancient and available variety of Nothing, with traditional functions in artistic representation. It is a terminus, a distancing, suggesting an ultimate perspective or perhaps a last judgement; and its cold reality can be used to spook the complacent acceptance of a here-and-now to which listeners are inevitably committed.

One of our aims is to right this neglect of nothing and show a little of the curious way in which Nothing in all its guises has proved to be a key concept in many human inquiries, whose right conception has opened up new ways of thinking about the world. We will begin our nullophilia by investigating the history of the concept and symbol for the mathematicians' zero. Here, nothing turns out to be quite as one expected. The logic of the Greeks prevents them having the idea at all and it is to the Indian cultures that we must look to find thinkers who are comfortable with the idea that Nothing might be something. Next, we shall follow what happened after the Greeks caught up. Their battle with zero focused upon its manifestation as a physical zero, the zero of empty space, the vacuum and the void. The struggle to make sense of these concepts, to incorporate them into a cosmological framework that impinged upon everyday experiences with real materials, formed the starting point for an argument that would continue unabated, becoming ever more sophisticated, for nearly two thousand years. Medieval science and theology grappled constantly with the idea of the vacuum, trying to decide questions about its physical reality, its logical possibility and its theological desirability.

Part of the problem with zero, as with the complementary concept of infinity, was the way in which it seemed to invite paradox and confusing self-reference. This was why so many careful thinkers had given it such a wide berth. But what was heresy to the logician was a godsend to the writer. Countless authors avoided trouble with Nothing by turning over its paradoxes and puns, again and again, in new guises, to entertain and perplex. Whereas the philosopher might face the brunt of theological criticism for daring to take such a sacrilegious concept seriously, the humorist trying to tell his readers that 'Nothing really matters' could have his cake and eat it, just as easily as Freddie Mercury. If others disapproved of Nothing, then the writer's puns and paradoxes just provided more ammunition to undermine the coherence of Nothing as a sensible concept. But when it came back into fashion amongst serious thinkers, then were not his word games profound explorations of the bottomless philosophical concept that Nothingness presented?

Hand-in-hand with the searches for the meaning of Nothing and the void in the Middle Ages, there grew up a serious experimental philosophy of the vacuum. Playing with words to decide whether or not a vacuum could truly exist was not enough. There was another route to knowledge. See if you could make a vacuum. Gradually, theological disputes about the reality of a vacuum became bound up with a host of simple experiments designed to decide whether or not it was possible to evacuate a region of space completely. This line of inquiry eventually stimulated scientists like Torricelli, Galileo, Pascal and Boyle to use pumps to remove air from glass containers and demonstrate the reality of the pressure and weight of the air above our heads. The vacuum had become part of experimental science. It was also very useful.

Still, physicists doubted whether a true vacuum was possible. The Universe was imagined to contain an ocean of ethereal material through which we moved but upon which we could exert no discernible effect. The science of the eighteenth and nineteenth centuries grappled with this elusive fluid and sought to use its imagined presence to explain the newly appreciated natural forces of electricity and magnetism. It would only be banished by Einstein's incisive genius and Albert Michelson's experimental skill. Together they removed the need and the evidence for a cosmic ether. By 1905 a cosmic vacuum had become possible again.

Things soon changed. Einstein's creation of a new and spectacular theory of gravity allowed us to describe a space that is empty of mass and energy with complete mathematical precision. Empty universes could exist.

Yet something had been missed out in the world of the very small. The quantum revolution showed us why the old picture of a vacuum as an empty box was untenable. Henceforth, the vacuum was simply the state that remained when everything that could be removed from the box was removed. That state was by no means empty. It was merely the lowest energy state available. Any small disturbances or attempts to intervene would raise its energy.

Gradually, this exotic new picture of quantum nothingness succumbed to experimental exploration. The multiplication of artificial voids by scientists at the end of the nineteenth century had paved the way for all sorts of useful and now familiar developments in the form of vacuum tubes, light bulbs and X-rays. Now the 'empty' space itself started to be probed. Physicists discovered that their defensive definition of the vacuum as what was left when everything that could be removed had been removed was not as silly as it sounds. There *was* always something left: a vacuum energy that permeated every fibre of the Universe. This ubiquitous, irremovable vacuum energy was detected and shown to have a tangible physical presence. Only relatively recently has its true importance in the cosmic scheme of things begun to be appreciated. We shall see that the world may possess many different vacuum states. A change from one to another may be possible under certain circumstances, with spectacular results. Remarkably, it appears that such a transition is very difficult to avoid during the first moments of our Universe's expansion. More remarkable still, such a transition could have a host of nice consequences, showing us why the Universe possesses many unusual properties which would otherwise be a complete mystery to us.

Finally, we shall run up against two cosmological mysteries about Nothing. The first is ancient: the problem of creation out of nothing – did the Universe have a beginning? If so, out of what did it emerge? What are the religious origins of such an idea and what is its scientific status today? The second is modern. It draws together all the modern manifestations of the vacuum, the description of gravity and the inevitability of energy in a quantum vacuum. Einstein showed us that the Universe might contain a mysterious form of vacuum energy. Until very recently, astronomical observations could only show that if this energy is present, as an all-pervading cosmic influence, then its intensity must be fantastically small if it is not to come to dominate everything else in the Universe. Physicists have no idea how its influence could remain so small. The obvious conclusion is that it isn't there at all. There must be some simple law of Nature that we have yet to find that restores the vacuum and sets this vacuum energy equal to zero. Alas, such a hope may be forlorn. Last year, two teams of astronomers used Earth's most powerful telescopes together with the incomparable optical power of the Hubble Space Telescope to gather persuasive evidence for the reality of the cosmic vacuum energy. Its effects are dramatic. It is accelerating the expansion of the Universe. And if its presence is real, it will set the future course of the Universe, and determine its end. What better place to begin?

chapter one

Zero – The Whole Story

"Is it not mysterious that we can know more about things which do not exist than about things which do exist?"

Alfréd Renyi[1]

"Round numbers are always false."

Samuel Johnson[2]

THE ORIGIN OF ZERO

"The great mystery of zero is that it escaped even the Greeks."

Robert Logan[3]

When we look back at the system of counting that we learned first at school it seems that the zero is the easiest bit. We used it to record what happens when nothing is left, as with a sum like 6 minus 6, and anything that gets multiplied by zero gets reduced to zero, as with $5 \times 0 = 0$. But we also used it when writing numbers to signal that there is an empty entry, as when we write one-hundred and one as 101.

These are such simple things – much simpler than long division, Pythagoras' Theorem, or algebra – that it would be easy to assume that zero must have been one of the first pieces of arithmetic to be developed by everyone with a counting system, while the more difficult ideas like geometry and algebra were only hit upon by the most sophisticated cultures. But this would be quite wrong. The ancient Greeks, who developed the logic and geometry that form the basis for all of modern mathematics, never introduced the zero symbol. They were deeply suspicious of the whole idea. Only three civilisations used the zero, each of them far from the cultures that would evolve into the so-called Western world, and each viewed its role and meaning in very different ways. So why was it so difficult for the zero symbol to emerge in the West? And what did the difficulty have to do with Nothing?

As the end of the year 1999 approached, the newspapers devoted more and more copy to the impending doom that was to be wrought by the Millennium Bug. The reason for this collective loss of sleep, money and confidence was the symbol 'zero', or two of them to be more precise. When the computer programs that control our transport and banking systems were first written, computers were frugal with memory space – it was much more expensive than it is today.[4] Anything that could save space was a money-saving bonus. So when it came to dating everything that the computer did, instead of storing, say, 1965, the computer would just store the last two digits, 65. Nobody thought as far ahead as the year 2000 when computers would be faced with making sense of the truncated 'date' 00. But if there is one thing that computers really don't like, it's ambiguity. What does 00 mean to the computer? To us it's obviously short for year 2000. But the computer doesn't know it isn't short for 1900, or 1800 for that matter. Suddenly, you might be told that your credit card with its 00 expiry year is 99 years out of date. Born in 1905? Maybe the computer would soon be mailing out your new elementary-school application forms. Still, things didn't turn out as badly as the pessimists predicted.[5]

Counting is one of those arts, like reading, into which we are thrust during our first days at school. Humanity learned the same lessons, but took thousands of years to do it. Yet whereas human languages exist by the thousand, their distinctiveness often enthusiastically promoted as a vibrant symbol of national identity and influence, counting has come to be a true human universal. After the plethora of our languages and scripts for writing them down, a present-day tourist from a neighbouring star would probably be pleasantly surprised by the complete uniformity of our systems of reckoning. The number system looks the same everywhere: ten numerals – 1, 2, 3, 4, 5, 6, 7, 8, 9 and 0 – and a simple system that allows you to represent any quantity you wish: a universal language of symbols. The words that describe them may differ from language to language but the symbols stay the same. Numbers are humanity's greatest shared experience.

The most obvious defining feature of our system of counting is its use of a base of ten. We count in tens. Ten ones make ten; ten tens make one hundred; and so on. This choice of base was made by many cultures and its source is clearly to be found close at hand with our ten fingers, the first counters. Sometimes one finds this base is mixed in with uses of 20 as a base (fingers plus toes) in more advanced cultures, whilst less advanced counting systems might make use of a base of two or five.[6] The exceptions are so rare as to be worth mentioning. In America one finds an Indian counting system based on a base of eight. At first this seems very odd, until you realise that they were also finger counters – it is just that they counted the eight gaps between the fingers instead of the ten fingers.

You don't have to be a historian of mathematics to realise that there have been other systems of numbers in use at different times in the past. We can still detect traces of systems of counting that differ in some respects from the decimal pattern. We measure time in sets of 60, with 60 seconds in a minute, 60 minutes in an hour, and this convention is carried over to the measurement of angles, as on a protractor or a navigator's compass. Elsewhere, there are relics of counting in twenties:[7] 'three-score years and ten' is the expected human lifetime, whilst in French the number words for 80 and 90 are *quatre-vingts* and *quatre-vingt-dix*, that is four-twenties and four-twenties and ten. In the commercial world we often order by the gross or the dozen, witness to a system with a base of twelve somewhere in the past.

The ten numerals 0, 1, . . . , 9 are used everywhere, but one other system for writing numbers is still in evidence around us. Roman numerals are often to be found on occasions where we want to emphasise something dynastic, like Henry VIII, or traditional, like the numbers on the clock face in the town square. Yet Roman numerals are rather different from those we use for arithmetic. There is no zero sign. And the information stored in the symbols is different as well. Write III and we interpret it as one hundred plus one ten plus one: one-hundred and eleven. Yet to Julius Caesar the marks III would mean one and one and one: three. These two missing ingredients, the zero sign and a positional significance when reading the value of a symbol, are features that lie at the heart of the development of efficient human counting systems.

EGYPT – IN NEED OF NOTHING

"Joseph gathered corn as the sand of the sea, very much, until he left numbering; for it was without number."

Genesis 41

The oldest developed counting systems are those used in ancient Egypt and by the Sumerians in Southern Babylonia, in what is now Iraq, as early as 3000 BC. The earliest Egyptian hieroglyphic[8] system used the repetition of a suite of symbols for one, ten, a hundred, a thousand, ten thousand, a hundred thousand and a million. The symbols are shown in Figure I.I. The

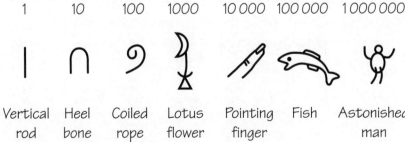

1	10	100	1000	10 000	100 000	1 000 000
Vertical rod	Heel bone	Coiled rope	Lotus flower	Pointing finger	Fish	Astonished man

Figure I.I *Egyptian hieroglyphic numerals.*

Egyptian symbols for the numerals one to nine are very simple and consist of the repetition of an appropriate number of marks of the vertical stroke, |, the symbol for one; so three is just |||. The symbols for the larger multiples of ten are more picturesque. Ten is denoted by an inverted u, a hundred by a coil, a thousand by a lotus flower, ten thousand by a bent finger, a hundred thousand by a frog or a tadpole with a tail, and a million by a man with his arms raised to the heavens. With the exception of the sign for one, they seem to have no obvious connection with the quantities they denote. Some connections are probably phonetic, deriving from the similar sounds for the things pictured and the original words used to describe the quantities. Only the bent finger marking ten thousand seems to hark back to a system of finger counting. We can only guess about the others. Perhaps the tadpoles were so numerous in the Nile when the frogs' spawn hatched in the spring that they symbolised a huge number; maybe a million was just an awesomely large quantity, like the populations of stars in the heavens above.

The symbols were written differently if they were to be read from right to left or left to right in an inscription.

Hieroglyphs were generally written down from right to left so that our number 3,225,578 would appear as shown in Figure 1.2.

One of the oldest examples of these numerals appears on the handle of a club belonging to King Narmer, who lived in the period 3000–2900 BC, celebrating the fact that the loot seized in one of his military campaigns amounted to 400,000 bulls, 1,422,000 goats and 120,000 human prisoners. The symbols for these quantities beneath pictures of a bull, a goat and a seated figure can be seen on the bottom right of Figure 1.3. The order in

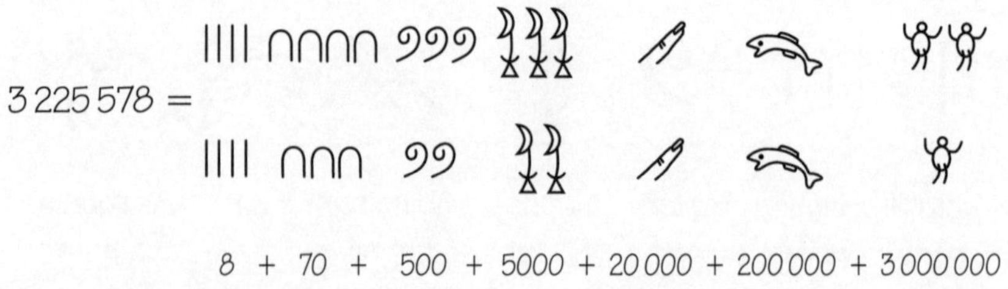

$$3\,225\,578 =$$

$$8 \;+\; 70 \;+\; 500 \;+\; 5000 \;+\; 20\,000 \;+\; 200\,000 \;+\; 3\,000\,000$$

Figure 1.2 *The hieroglyph for our number three million, two hundred and twenty-five thousand, five hundred and seventy-eight.*

which the symbols are written is not important because there are different symbols for one, ten and a hundred. The hieroglyph

$$\cap\cap|||$$

would signal exactly the same quantity if written forwards or backwards. The symbols can be laid out in any way at all without changing the value of the number they are representing. However, Egyptian stonemasons were given strict rules of style for writing numbers: signs were to appear from right to left in descending order of size on a line underneath the symbol for the object that was being counted (as in Figure I.3). However, there was a tendency to group similar symbols together over two or three lines to help the reader quickly read off the total, as shown in Figure I.4.

Thus we see that the relative positions of the Egyptian counting symbols carry no numerical information and so there is no need for a symbol for zero. When the number symbols can sit in any location without altering the total quantity they are representing, there is no possibility of an empty 'slot' and no meaning to a signal of its presence. The need for a zero arises when you have nothing to count – but in that case you write no symbols at

Figure I.3 *Hieroglyphics inscribed on the handle of King Narmer's war club,*[9] *3000–2900 BC.*

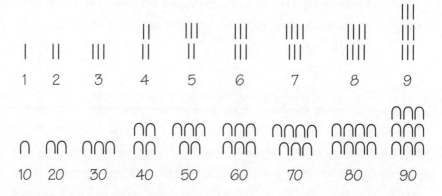

Figure I.4 *The grouping of number signs to help the reader.*

all. The Egyptian system is an early example of a decimal system (the collective unit is 10) with symbols for numbers which carry no positional information. In such a system there is no place for a zero symbol.

BABYLON – THE WRITING IS ON THE WALL

"In the same hour came forth fingers of a man's hand, and wrote over against the candlestick upon the plaster of the wall of the king's palace … And this is the writing that was written, Mene, Mene, Tekel, Upharsin. This is the interpretation. Mene; God hath numbered thy kingdom and finished it. Tekel; Thou art weighed in the balances, and art found wanting. Peres; thy kingdom is divided.'

Daniel 5[10]

The earliest Sumerian system, also in use around 3000 BC, was more complex than that employed by the Egyptians and seems to have developed independently. It was later adopted by the Babylonians and so the two civilisations are usually regarded as different parts of a single cultural development. The motivation for their systems of writing and counting was at first administrative and economic. They kept detailed records and accounts of exchanges, stores and wages. Often, a detailed list of items will be found on one side of a tablet, with the total inscribed on the reverse.

The counting system of early Sumer was not solely decimal. It made good use of the base ten to label quantities but it also introduced 60 as a second base number.[11]

It is from this ancient system that we inherited our pattern of time-keeping with 60 seconds to the minute and 60 minutes to the hour. Expressing 10 hours 10 minutes and 10 seconds in seconds shows us how to unfold a base-60 counting system. We have a total of $(10 \times 60 \times 60) + (10 \times 60) + 10 = 36,000 + 600 + 10 = 36,610$ seconds.

The Sumerians had number words for the quantities 1, 60, 60×60, $60 \times 60 \times 60$, ... and so on. They also had words for the numbers 2, 3, 4, 5, 6, 7, 8, 9 and 10, together with the multiples of ten below 60. A distinct word was used for 20 (unrelated to the words for 2 and 10) but 'thirty' was a compound word meaning 'three tens', 'forty' meant 'two twenties', and 'fifty' meant 'forty and ten'. So there was a weaving of base 10 and base 20 elements to ease the jump up from one to sixty.

Whereas the Egyptians carved their signs in stone with hammers and chisels or painted them on to papyri with reeds, Sumerian records were kept by making marks in tablets of wet clay. Stone was not common in Sumer and other media like papyrus or wood would rapidly perish or rot, but clay was readily available. The inscriptions were made by impressing the wet clay with two types of reed or ivory stylus, shaped like pencils of differing widths. The round blunt end allowed notches or circular shapes to be impressed whilst the sharp end allowed lines to be drawn. The sharp end was used for writing whilst the blunt end was used for representing numbers. The original symbols are shown in Figure 1.5 and are called *curvilinear* signs. The number symbols[12] usually appeared over an image of the thing being enumerated and reveal a new feature, not present in Egypt. The symbol for 600 combines the large notch, representing 60, with the small circle, representing 10. Likewise, the symbol for 36,000 combines the large circle, for 3,600, with the small circle, for 10. This economical scheme creates a multiplicative notation. There are fewer symbols to learn and the symbols for large numbers have an internal logic that enables larger numbers to be generated from smaller ones

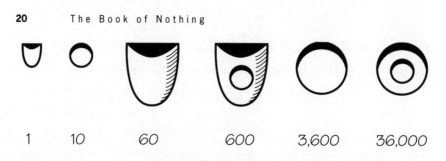

| 1 | 10 | 60 | 600 | 3,600 | 36,000 |

Figure 1.5 *The impressed shapes representing Sumerian numerals on clay tablets.*

without inventing new symbols. However, notice that you have to do a little bit of mental arithmetic every time you want to read a large number! The system is additive and there is again no significance to the positions of the symbols when they are inscribed on the clay tablets. As in Egypt, similar symbols were grouped together for stylistic reasons and for ease of reckoning. The early style was to gather marks into pairs. For example, the decimal number 4980 is broken down as

$$4980 = 3600 + 1380$$
$$= 3600 + 600 + 600 + 60 + 60 + 60$$

and this would be written as shown in Figure 1.6 since tablets were read from right to left and from top to bottom.

A tedious feature of this system is the huge number of marks that have to be made in order to represent large numbers that are not exact multiples of 60. To overcome this problem, scribes developed a shorthand subtraction notation, introducing a 'wing' sign that played the role of our minus sign so that they could write a number like 59 as 60 minus 1 by means of the three symbols (Figure 1.7) instead of the fourteen marks that would otherwise have been required.[13]

By 2600 BC a significant change had occurred in the way that the Sumerian number characters were written. The reason: new technology – in the form of a change of writing implement. A wedge-shaped stylus was introduced which could produce sharper lines and wedge-shapes of different sizes. These became known as 'cuneiform'[14] signs and only two marks are used, a vertical wedge denoting 'one' and a chevron representing 'ten'

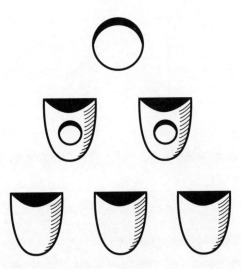

Figure I.6 *The number 4980 in early Sumerian representation, before 2700 BC.*

(Figure I.8). Again, the fusion of symbols can be used to build up large numbers from smaller ones. If the symbols for 60 and 10 were in contact they signified a multiplication of values (600) whereas if they were separated they signified an addition (70). However, some care was needed to make sure that juxtapositions of signs like these did not become confused. The Sumerian combinations of symbols avoided this problem because the individual marks were much more distinct.

Another problem was the distinction of the signs for 1 and for 60. Their shapes are identical wedges and at first they were distinguished simply by making the 60 wedge bigger. Later, it was done by separating the

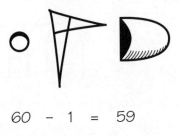

60 – 1 = 59

Figure I.7 *The number 59 written as 60 minus 1.*

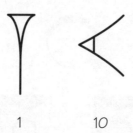

1   10

Figure 1.8 *The cuneiform impressions made by the two ends of the scribe's stylus, denoting the numbers 1 and 10.*

wedge shape for 60 from those for numbers less than nine. The writing of the number 63 is shown in Figure 1.9.

Many other systems of counting can be found around the ancient world which use the same general principles as these. The Aztecs (AD 1200) had an additive base-20 system with symbols for 1, 20, 400 = 20 × 20, and 8000 = 20 × 20 × 20. The Greeks (500 BC) used a base-10 system with different signs for 1, 10, 100, 1000, 10,000 but supplemented them with a further sign for 5 which they then added to the other signs to generate new symbols for 50, 500, 5000, and so on (see Figure 1.10).

All these systems of writing numbers are cumbersome and laborious to use if you want to do calculations that involve multiplication or division. The notation does not do any work for you, it is just like a shorthand for

(a)

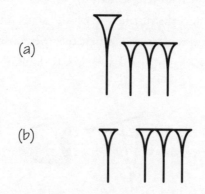

(b)

Figure 1.9 *Two ways of writing the number 63: (a) using a larger version of the 60 symbol to separate 63 as 60 and 3, or (b) by leaving a space between the symbol for 60 and those for 3.*

Ι Γ Δ Γ⁴ Η Γᴾ Χ Γˣ Μ

1   5   10   50   100   500   1000   5000   10,000

6668 is 5000 + 1000 + 500 + 100 + 50 + 10 + 5 + 3

Γˣ Χ Γᴾ Η Γ⁴ Δ Γ ΙΙΙ

Figure I.10 *Greek numerals, which first appeared around 500 BC, used combinations of symbols to generate higher numbers. As an example, we have written the number 6668.*

writing down the number words in full. The next step in sophistication, a step that was to culminate in the need to invent the zero symbol, was to introduce a *positional* or *place value* system in which the locations of symbols determined their values. This allows fewer symbols to be used because the same symbol can have different meanings in different locations or when used in different contexts.

A positional system appeared first in Babylonia around 2000 BC. It simply extended the cuneiform notation and the old additive base-60 system to include positional information. It was used by mathematicians and astronomers rather than for everyday accounting because the old system allowed the reader to see the relative sizes of numbers more easily. Many scribes must therefore have practised with both systems. However, it was used in the recording of royal decrees and so must have been understood by a broad cross-section of the Babylonian public. Thus, a number like 10,292 would be conceived in our notation as $[2; 51; 32] = (2 \times 60 \times 60) + (51 \times 60) + 32$, and written in cuneiform as shown in Figure I.11. This is just like our representation of a number like 123 as $(1 \times 10 \times 10) + (2 \times 10) + 3$. Our notation just reads off the number that multiplies the number of contributions by each power of 10. We still retain the Babylonian system for time measures. Seven hours and five minutes and six seconds is just $(7 \times 60 \times 60) + (5 \times 60) + 6 = 25,506$ seconds.

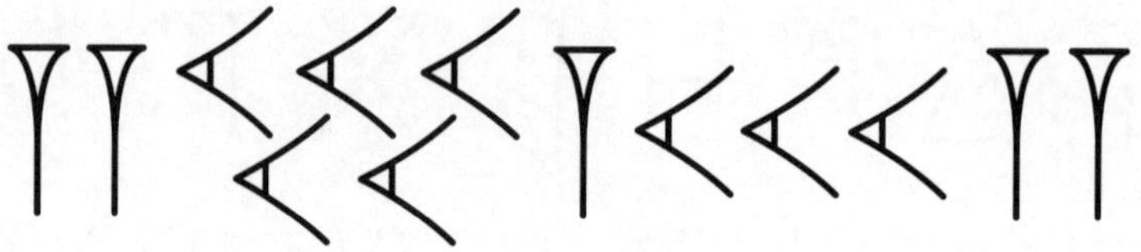

Figure I.11 *The number 10,292 in cuneiform.*

The earliest positional decimal system like our own did not appear until about 200 BC when the Chinese introduced the place value system into their base-10 system of signs. Their rod number symbols, together with an example of their positional notation in action, are shown in Figure I.12.

THE NO-ENTRY PROBLEM AND THE BABYLONIAN ZERO

"There aren't enough small numbers to meet the many demands made of them."

Richard K. Guy[15]

These advances were not without their problems. The Babylonian system was really a hybrid of positional and additive systems because the marking of the number of each power of 60 was still denoted in an additive fashion. This could produce ambiguity if sufficient space was not left between one order of 60 and the next. For instance, the symbols for 610 = [10; 10] = (10 × 60) + 10 could easily be misread for 10 + 10 (see Figure I.13). This was generally dealt with by separating the different orders of 60 clearly. Eventually, a *separation marker* was introduced to make the divisions unambiguous. It consisted of two wedge marks, one on top of the other, as shown in Figure I.14.

Any difficulties of interpretation would be compounded further if there was no entry at all in one of the orders. The spacing would then be more tricky to interpret. Imagine that our system had no 0 symbol and relied on careful spacing to distinguish 72 (seventy-two) from 7 2 (seven hundred and two). With different writing styles to contend with there would be many problems which are exacerbated if one has to distinguish 7

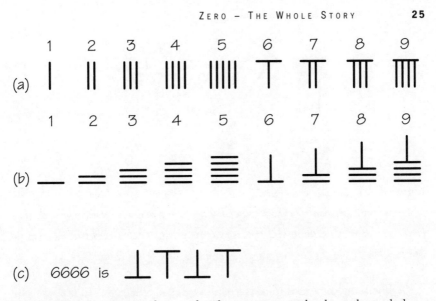

Figure 1.12 (a) *Chinese rod numerals. They are pictures of bamboo or bone calculating rods. When these symbols were used in the tens or thousands position they were rotated, and written as in (b), so our number 6666 would have been expressed as shown in (c).*

2 (seven thousand and two) as well as 7 2 and 72. The more spaces that you need to leave, the harder it becomes to judge.[16] This is why positional notation systems eventually need to invent a *zero* symbol to mark an empty slot in their positional representation of a number. The more sophisticated their commercial systems the greater is the pressure to do so. For nearly 1500 years the Babylonians worked without a symbol for 'no entry' in their register of different powers of ten or sixty; they merely left a space. Their success required a good feeling for the magnitudes of the astronomical and mathematical problems they were dealing with, so that large discrepancies from expected answers could be readily detected.

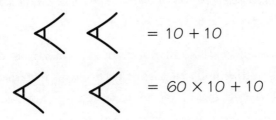

Figure 1.13 *The Babylonian forms for 610 and 20 could easily be confused.*

$$2 \times 60 \times 60 + 0 \times 60 + 15$$

Figure I.14 *The Babylonians first introduced a 'separator' symbol to mark empty spaces in the number expression. They were shaped like double chevrons and created by two overlapping impressions of the stylus wedge. This example was found on a tablet recording astronomical observations, dated between the late third and early second century BC.*

The Babylonian solution to the no-entry problem was to use a variant of the old separation marker sign to signal that there was no entry in a particular position. This appears in writing of the fourth century BC, but may have been in existence for a century earlier because of the paucity of earlier documents and the likelihood that some of those that do exist are copies of earlier originals. A typical example of the use of the Babylonian zero is shown in Figure I.15, where the number $3612 = 1 \times (60 \times 60) + (0 \times 60) + (1 \times 10) + 2$ is written:

$$1 \quad ; \quad 0 \quad ; \quad 10 \quad ; \quad 2$$

Figure I.15 *An example of the Babylonian zero as used in the third or second century BC to write $3612 = (1 \times 60 \times 60) + (0 \times 60) + (1 \times 10) + 2$.*

Babylonian astronomers[17] also made extensive use of zero at the end of a character string and we find examples of 60 being distinguished from 1 by writing it as shown in Figure I.16. So we begin to see how the Babylonian zero functioned in a similar way to our own. It began, like the positional notation, as shorthand used by Babylonian mathematicians. This ensured

$= (1 \times 60) + 0 = 60$

Figure 1.16 *The number sixty in an astronomical record was also used at the end of strings of numerals, as shown here.*

its extensive use by Babylonian astronomers, and it is because of the huge importance and persistence of Babylonian astronomy that their system of counting remained so influential as the centuries passed.

This is the culmination of the Babylonian development: the first symbolic representation of zero in human culture. In retrospect, it seems such a straightforward addition to their system that it is puzzling why it took more than fifteen centuries to pass from the key step of a positional notation to a system with an explicit zero symbol.

Yet the Babylonian zero should not be identified totally with our own. For the scribes who etched the double chevron sign on their clay tablets, those symbols meant nothing more than an 'empty space' in the accounting register. There were no other shades of meaning to the Babylonian 'nothing'. Their zero sign was never written as the answer to a sum like 6 – 6. It was never used to express an endpoint of an operation where nothing remains. Such an endpoint was always explained in words. Nor did the Babylonian zero find itself entwined with metaphysical notions of nothingness. There is a total absence of any abstract interweaving[18] of the numerical with the numinous. They were very good accountants.

THE MAYAN ZERO

"I have nothing to say
and am saying it and that is
poetry."

John Cage[19]

The third invention of the positional system occurred in the remarkable Mayan culture that existed from about AD 500 until 925. Paradoxically, despite achieving great sophistication in architecture, sculpture, art, road building, writing, numerical calculation, calendar systems and predictive astronomy, the Mayans never invented the wheel, never discovered metal or glass, had no clocks which could measure time intervals of less than a day, and never made use of beasts of burden. Stone Age practices went hand in hand with extraordinary arithmetical sophistication. Why their culture ended so suddenly is still a mystery. All that remains are abandoned cities in the jungles and grasslands of present-day Mexico, Belize, Honduras and Guatemala. All manner of disasters have been suggested for the exodus of the population. Plague or civil war or earthquakes have all been blamed. A better bet is agricultural exhaustion of their soil through persistent intensive farming and overuse.

The Mayan counting system was founded upon a base of 20 (see Figure I.17) and the numbers were composed of combinations of dots (each denoting 'one') and rods (each denoting 'five'). The first nineteen numbers were built up with dots and lines in a simple additive fashion, probably derived from an earlier finger-and-toe counting system.[20] The dot (or sometimes a small circle) used as a symbol for 'one' is found throughout the Central American region at early times and was probably linked to the use of cocoa beans as a currency unit. As in the Babylonian culture, there was a distinction between everyday calculation and the higher computations of mathematicians and astronomers.

When one needed to write numbers larger than 20 a tower of symbols was created, the bottom floor marking multiples of I, the first floor multiples of 20. However, the second floor did *not* read multiples of 20 ×

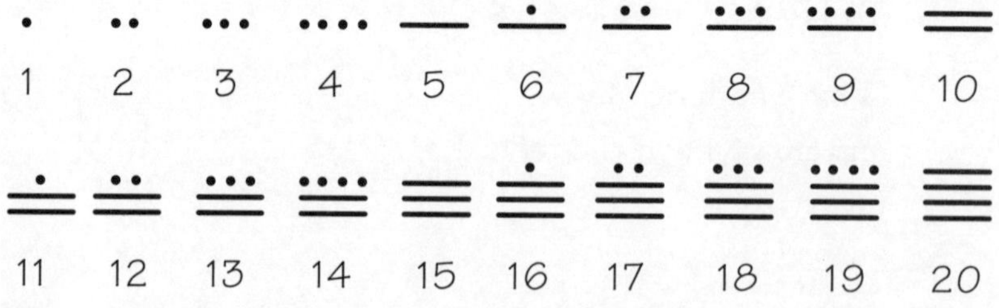

Figure I.17 *The numbers from 1 to 20 in the Mayan system used by priests and astronomers.*

20. It carried multiples of 360! But the pattern then carried on unbroken. The next level up then carried multiples of $20 \times 360 = 7200$; then $20 \times 7200 = 144{,}000$ and all subsequent levels were each 20 times the level below. Numbers were read downwards. The number $4032 = (11 \times 360) + (3 \times 20) + 12$ is shown in Figure 1.18.

Thus we see that the Mayans had a positional, or place-value, system and to this they added a symbol for zero, to denote no entry on one of the levels of the number tower. The symbol they used is very curious. It resembles a shell or even an eye, comes in a number of slightly different forms, and seems to have conveyed the idea of completion, reflecting its aesthetic role in representing the numbers which we will describe below. Some of the zero shapes are shown in Figure 1.19. Thus the number $400 = (1 \times 360) + (2 \times 20) + 0$ would be written as shown in Figure 1.20. The Mayans used their zero symbol in both intermediate and final positions in their symbol strings, just as we do.

The curious step in the Mayan system at level two, marked by 360 rather than 400 as would have been characteristic of a pure base-20 system, means that the zero symbol differs from our own in one very important respect. If we add a zero symbol to the right-hand end of any number then we multiply its value by 10, the value of our system's base; thus $170 = 17 \times 10$. If a counting system of any base proceeds through levels which are each related to the previous one by a power of the base, whatever its value, then adding a zero to a symbol string will always have the effect of multiplying the number by the base value. The Mayan system lacked this nice property because of the uneven steps from level to level. It stopped the Mayans from exploiting their system to the full.

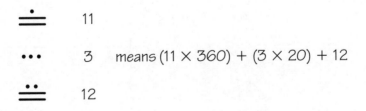

means $(11 \times 360) + (3 \times 20) + 12$

Figure 1.18 *The Mayan representation of the number 4032.*
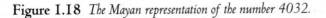

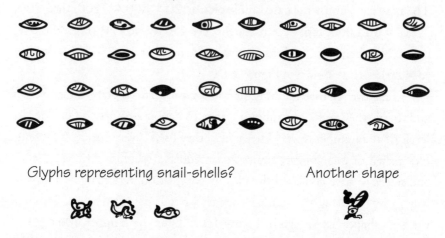

Glyphs representing sea-shells?

Glyphs representing snail-shells?   Another shape

Figure I.19 *Different symbolic forms for the Mayan zero (see note 9). They look like the shells of snails and sea creatures, or human eyes.*

The Mayans failed to introduce an even sequence of levels for a reason; they had other jobs for their counting system. It was designed to play a particular role keeping track of their elaborate cyclic calendar. They had three types of calendar. One was based upon a sacred cycle of 260 days, the *tzolkin*, which was split into 20 periods of 13 days. The second was a civil 'year' of 365 days, called the *haab*, which was divided into 18 periods of 20 days each plus a transition period of 5 days. The third calendar was based on a period of 360 days, called the *tun*, which was divided into eighteen periods of 20 days. Twenty *tun* equalled one *katun* (*ka* was the word for 20); twenty *katuns* was one *baktun* (*bak* was the word for 20 × 20); one *uinal* equalled 20 days.[21] Special hieroglyphs were used to represent these peri-

$$
\left.
\begin{array}{l}
1 \times 360 \\
+ \\
2 \times 20 \\
+ \\
0
\end{array}
\right\} = 400
$$

Figure I.20 *The Mayan representation of 400.*

Figure 1.21 *A Mayan hieroglyph denoting a length of time. For each of the units, baktun, katun, uinal and day, a special picture was used, usually of a head with other defining features or adornments. Alongside each picture was a numeral, composed of dots and bars, to indicate how many of those units should be taken. Sometimes small numbers, requiring only two dots or bars would have further ornaments added to balance the space. Here, reading from left to right and top to bottom, we have a representation of* 9 baktun *and* 14 katun *and* 12 tun *and* 4 uinal *and* 17 kin. *This gives a total* 3892 tun *and* 97 kin, *or* 1,401,217 kin *(days).*

ods. A complete picture denoting a period of time would then combine symbols for the time intervals with those signifying how many multiples of them were meant. The hieroglyph in Figure 1.21 should be read from left to right and from top to bottom and records the following times: 9 *baktun,* 14 *katun,* 12 *tun,* 4 *uinal* and 17 *kin* (days).

   In these pictograms the zero was represented by a number of exotic glyphs,[22] a few of which are shown in Figure 1.22.

Figure 1.22 *The various hieroglyphs for zero found on Mayan columns and statues.*

In this scheme the zero symbol is not essential for recording dates. What is novel about the Mayan zero is that it was introduced for aesthetic reasons. Without the zero picture, the pictogram for a date would have had a vacant patch and would look unbalanced. The elaborate zero glyphs filled the gap and created a dramatic rendering of a date which reinforced the religious significance of the numbers being represented.

THE INDIAN ZERO

"The Indian zero stood for emptiness or absence, but also space, the firmament, the celestial vault, the atmosphere and ether, as well as nothing, the quantity not to be taken into account, the insignificant element."

Georges Ifrah[23]

The destruction of the Babylonian and Mayan civilisations prevented their independent inventions of the zero symbol from determining the future pattern of representation. That honour was to be given to the third inventor of the zero whose way of writing all numbers is still used universally today.

The Hindus of the Indus valley region had a well-developed culture as early as 3000 BC. Extensive towns were established with water systems and ornaments. Seals, writing systems and evidence of calculation witness to a sophisticated society. Writing and calculation spread throughout the Indian sub-continent over the following millennia. A rich diversity of calligraphic styles and numeral systems can be found throughout Central India and in nearby regions of South-East Asia which made use of the Brahmi numerals. This notation appeared for the first time about 350 BC, although only examples of the numerals 1, 2, 4 and 6 still remain on stone monuments. Transcriptions in the first and second century BC show what they probably looked like[24] (see Figure 1.23).

The forms of the Brahmi numerals are still something of a mystery. The signs for the numerals from 4 to 9 do not have any obvious association with the quantities they denote, but they may derive from alphabets that

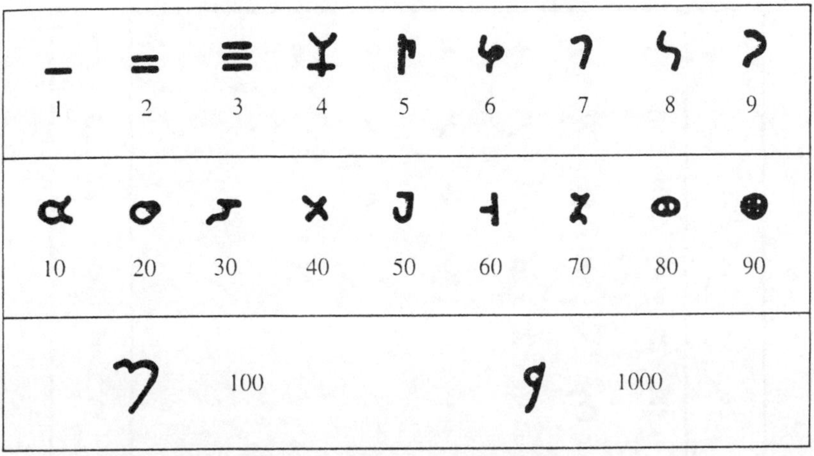

—	=	☰	⅄	𝑃	𝞶	𝟽	⟍	⟩
1	2	3	4	5	6	7	8	9
α	σ	𝔧	✕	𝖩	⊣	𝘟	∞	⊕
10	20	30	40	50	60	70	80	90

⟩	100	𝑔	1000

Figure 1.23 *The early Indian symbols for the numerals 1 to 9.*

have disappeared or be an evolutionary step from an earlier system of numerals with clear interpretations that no longer exist.

The Brahmi system was transformed into a positional base-10, or decimal, notation in the sixth century AD. It exploited the existence of distinct numerals for the numbers 1 to 9 and a succinct notation for larger numbers and number words for the higher powers of ten. The earliest written example of its use goes back to AD 595 on a copperplate deed from Sankheda.[25]

The inspiration for this brilliant system is likely to have been the use of counting boards for laying out numbers with stones or seeds. If you want to lay out a number like 102 using stones, then place one stone in the hundreds column followed by a space in the tens column and a two in the units. A further motivation for devising a clear logical notation for dealing with very large numbers is known to have come from the studies of Indian astronomers, who were influenced by earlier Babylonian astronomical records and notations. The commonest positional notation emerging from the Brahmi numerals was that using the *Nâgarî* script, shown in Figure 1.24.

A unique feature of the Indian development of a positional system is the way in which it made use of the same numerals that were in existence

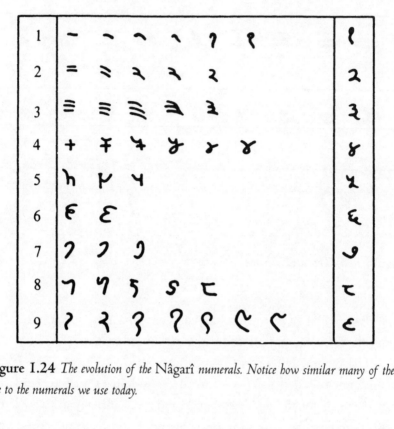

Figure I.24 *The evolution of the Nâgarî numerals. Notice how similar many of them are to the numerals we use today.*

long before. In other cultures the creation of a positional notation required a change of notation for the numerals themselves. The earliest known use of their place-value system is AD 594.

As we have learned from the Babylonians and the Mayans, once a positional system is introduced it is only a matter of time before a zero symbol follows. The earliest example of the use of the Indian zero is in AD 458, when it appeared in a surviving Jain work on cosmology, but indirect evidence indicates that it must have been in use as early as 200 BC. At first, it seems that it was denoted by a dot, rather than by a small circle. A sixth-century poem, *Vâsavadattâ*, speaks of how[26]

> "the stars shone forth . . . like zero dots . . . scattered in the sky."

Later, the familiar circular symbol, 0, replaced the dot and its influence spread east to China. It was used to mark the absence of an entry in any position (hundreds, tens, units) of a decimal number and, because the Indian decimal system was a regular one, with each level ten times the previous one, zero also acted as an operator. Thus, adding a zero to the end of a number string effected multiplication by 10 just as it does for us. A wonderful application of this principle is to be found in a piece of Sanskrit poetry[27] by Bihârîlâl in which he expresses his admiration for a beautiful woman by referring to the dot (*tilaka*) on her forehead[28] in a mathematical way:

> "The dot on her forehead
> Increases her beauty tenfold,
> Just as a zero dot [*ɔunya-binðu*]
> Increases a number tenfold."

Although the Indian zero was first introduced to mark an absent numeral in the same way as for the Babylonians and the Mayans, it rapidly assumed the status of another numeral. Also, in contrast to the other inventors of zero, the Indian calculators readily defined it to be the result of subtracting any number from itself. In AD 628, the Indian astronomer Brahmagupta defined zero in this way and spelled out the algebraic rules for adding, subtracting, multiplying and, most strikingly of all, dividing with it. For example,

> "When *ɔunya* is added to a number or subtracted from a number, the number remains unchanged; and a number multiplied by *ɔunya* becomes *ɔunya*."

Remarkably, he also defines infinity as the number that results from dividing any other number by zero and sets up a general system of rules for multiplying and dividing positive and negative quantities.

There have been some interesting speculations as to why the Indian zero sign assumed the circular shape.[29] After all, we have seen it assume a very different form in the Mayan and Babylonian scripts. Subhash Kak has proposed that it developed from the Brahmi symbol for ten. This resem-

Figure 1.25 *A possible separation of the fishlike symbol for 10 into a circle and a line representing 1, leaving the circle for a zero sign.*

bled a simple fish or a proportionality sign ∝. Later, in the first and second centuries, it looked like a circle with a 1 attached (see Figure 1.25). Hence, it is suggested that the symbol for ten may naturally have divided into the sign for 1, a single vertical stroke, and the remaining circle which had the zero value.

A fascinating feature of the zero symbol in India is the richness of the concept it represents. Whereas the Babylonian tradition had a one-dimensional approach to the zero symbol, seeing it simply as a sign for a vacant slot in an accountant's register, the Indian mind saw it as part of a wider philosophical spectrum of meanings for nothingness and the void. Here are some of the Indian words for zero.[30] Their number alone indicates the richness of the concept of nothing in Indian philosophy and the way in which different aspects of absence were seen to be something requiring a distinct label.[31]

<u>Word</u>	<u>Sanskrit Meaning</u>
Abhra	Atmosphere
Akâsha	Ether
Ambara	Atmosphere
Ananta	The immensity of space
Antariksha	Atmosphere
Bindu	A point
Gagana	The canopy of heaven
Jaladharapatha	Sea voyage
Kha	Space
Nabha	Sky, atmosphere
Nabhas	Sky, atmosphere
Pûrna	Complete

Randhra	Hole
Shûnya / sunya	Void
Vindu	Point
Vishnupada	Foot of Vishnu
Vyant	Sky
Vyoman	Sky or space

Bindu is used to describe the most insignificant geometrical object, a single point or a circle shrunk down to its centre where it has no finite extent. Literally, it signifies just a 'point', but it symbolises the essence of the Universe before it materialised into the solid world of appearances that we experience. It represents the uncreated Universe from which all things can be created. This creative potential was revealed by means of a simple analogy. For, by its motion, a single dot can generate lines, by whose motion can be generated planes, by whose motion can be generated all of three-dimensional space around us. The *bindu* was the Nothing from which everything could flow.

This conception of generation of something from Nothing led to the use of the *bindu* in a range of meditational diagrams. In the Tantric tradition the meditator must begin by contemplating the whole of space, before being led, shape by shape, towards a central convergence of lines at a focal point. The inverse meditational route can also be followed, beginning with the point and moving outwards to encompass everything, as in Figure 1.26, where the intricate geometrical constructions of the *Sriyantra* are created to focus the eye and the mind upon the convergent and divergent paths that link its central point to the great beyond.

The revealing thing we learn from the Indian conception of zero is that the *sunya* included such a wealth of concepts. Its literal meaning was 'empty' or 'void' but it embraced the notions of space, vacuousness, insignificance and non-being as well as worthlessness and absence. It possesses a nexus of complexity from which unpredictable associations could emerge without having to be subjected to a searching logical analysis to ascertain their coherence within a formal logical structure. In this sense the Indian development looks almost modern in its liberal free associations. At

Figure 1.26 *The* Sriyantra, *a geometric construction used as a meditational guide in parts of the Tantric tradition. The earliest known examples date from the seventh century* AD, *but simpler patterns date back to the twelfth century* BC. *It consists of an intricate nested pattern of triangles, polygons, circles and lines, converging upon a central point, or* bindu, *which was either the end or the beginning of the meditational development as it moved inwards or outwards through the patterns. Of the nine central triangles, four point upwards marking 'male' cosmic energy, and five point downwards marking 'female' energy. Considerable geometric knowledge was required to construct these and other Vedic guides to worship.*[32]

its heart is a specific numerical and notational function which it performs without seeking to constrain the other ways in which the idea can be used and extended. This is what we would expect to find in modern art and literature. An image or an idea may exist with a well-defined form and meaning in a specific science, yet be continually elaborated or reinvented by artists working with different aims and visions.

INDIAN CONCEPTIONS OF NOTHINGNESS

"It is true that as the empty voids and the dismal wilderness belong to zero, so the spirit of God and His light belong to the all-powerful One."

Gottfried Leibniz[33]

The Indian introduction of the zero symbol owes much to their ready accommodation of a variety of concepts of nothingness and emptiness. The Indian culture already possessed a rich array of different concepts of 'Nothing' that were in widespread use. The creation of a numeral to denote no quantity or an empty space in an accountant's ledger was a step that could be taken without the need for realignment of parts of any larger philosophy of the world. By contrast, the Hebrew tradition regarded the void as the state from which the world was created by the movement and word of God. It possessed a host of undesirable connotations. It was a state from which to recoil. It spoke of poverty and a lack of fruitfulness: it meant separation from God and the removal of His favour. It was anathema. Similarly, for the Greeks it was a serious philosophical dilemma. Their respect for logic led them into a quandary over the treatment of Nothing as if it were something.

The Indian religious traditions were more at home with these mystical concepts. Their religions accepted the concept of non-being on an equal footing with that of being. Like many other Eastern religions, the Indian culture regarded Nothing as a state from which one might have come and to which one might return — indeed these transitions might occur many times, without beginning and without end. Where Western religious traditions sought to flee from nothingness, the use of the dot symbol for zero in meditational exercises showed how a state of non-being was something to be actively sought by Buddhists and Hindus in order to achieve Nirvana: oneness with the Cosmos.

The hierarchy of Indian concepts of 'Nothing' forms a coherent whole. It includes the zero symbol of the mathematicians in an integrated way. In Figure 1.27,[34] the network of meanings gathered by Georges Ifrah is displayed. Notice how the network of meanings is linked to the ideas captured by the words for zero that we gave on pages 36–37. Amid this network of connected meanings, we begin to see some of the sources for our own multiple meanings for Nothing.

At the top level are words, including those which are associated with the sky and the great beyond. They are joined by *bindu*, reflecting its representation of the latent Universe. As we move down the tree we encounter a host of different terms for the absence of all sorts of properties: non-being, not formed, not produced, not created, together with another collection of terms that carry the meaning of being negligible, insignificant, or having no value.

These two separate threads of meaning merged in the abstract concept of zero so that, at least from the fifth century AD onwards, the concept of Nothing began to reflect all the facets of the early Indian nexus of Nothings, from the prosaic empty vessel to the mystics' states of non-being.

The Greek tradition was a complete contrast to that of the Far East. Beginning with the school of Thales, the Greeks placed logic at the pinnacle of human thinking. Their sceptical attitude towards the wielding of 'non-being' as some sort of 'something' that could be subject to logical development was exemplified by Parmenides' influential arguments against the concept of empty space. He maintained that all his predecessors, like Heraclitus, had been mistaken in adopting the view that all things (those of which we can say 'it is') were made of the same basic material, whilst at the same time speaking about empty space (that of which we can say 'it is not'). He maintained that you can only speak about what is: what is not cannot be thought of, and what cannot be thought of cannot be.

From this statement of the 'obvious', Parmenides believed that many conclusions followed, among them the theorem that empty space could not exist. But more unexpected was the further conclusion that neither time, motion nor change could exist either. Parmenides simply believed that whenever you think or speak you must think or speak about something and

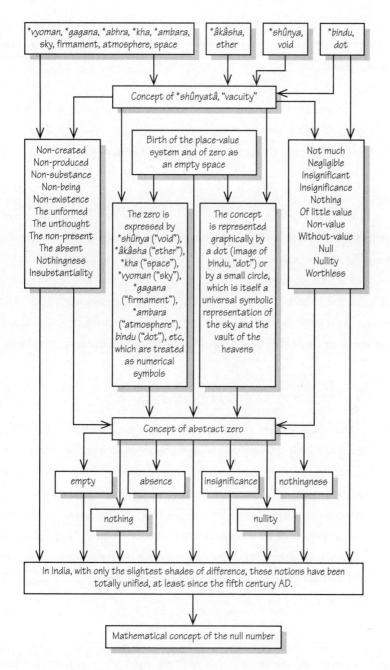

Figure I.27 *The array of interrelated meanings of the concepts associated with different aspects of nothingness in early Indian thought, culminating in the mathematical zero.*

so there must already exist real things to speak or think about. This implied that they must always have been there and can never change. Plato tells Theaetetus that

> "the great Parmenides ... constantly repeated in both prose and verse:
> Never let this thought prevail, that not being is,
> But keep your mind from this way of investigation."[35]

There are all sorts of problems with these ideas. How can Parmenides ever say that anything is not the case, or that something cannot be? Nevertheless, the legacy of his emphasis upon the need to be speaking about 'something' actual makes it very difficult to discuss concepts like the vacuum, Nothing, or even the zero of mathematics. From our vantage point this barrier seems strange. But whereas in India the zero could be introduced without straining any other philosophical position, in Greece it could not.

THE TRAVELLING ZEROS

> "The ingenious method of expressing every possible number using a set of ten symbols (each symbol having a place value and an absolute value) emerged in India. The idea seems so simple nowadays that its significance and profound importance is no longer appreciated ... The importance of this invention is more readily appreciated when one considers that it was beyond the two greatest men of antiquity, Archimedes and Apollonius."
>
> Pierre Simon de Laplace (1814)[36]

The Indian system of counting is probably the most successful intellectual innovation ever devised by human beings.[37] It has been universally adopted. It is found even in societies where the letters of the Phoenician alphabet are not used. It is the nearest thing we have to a universal language. Invariably, the result of trading contact between the Indian system of counting and any other system was for the former to be adopted by the latter or, at least, for its most powerful features to be imported into the local scheme. When the Chinese encountered the Indian system in the eighth century, they adopted the Indian circular zero symbol and a full place-value notation with nine numerals. The Indian system was introduced into Hebrew culture by the travelling scholar Ben Ezra (1092–1167), who journeyed widely in Asia and the Orient. He described the Indian system of counting in his influential *Book of Number*[38] and used the first nine letters of the Hebrew alphabet to represent the Indian numerals from 1 to 9 with a place-value notation but retained the small Indian circle to symbolise zero, naming[39] it after the Hebrew for 'wheel' (*galgal*). Remarkably, Ben Ezra single-handedly changed the old Hebrew number system into one with a place-value notation and a zero symbol, but there seemed to be no interest in his brilliant innovation and no one else took it up and developed it.

The Indian zero symbol found its way to Europe, primarily through Spain,[40] via the channel of Arab culture. The Arabs had close trading links with India which exposed them to the efficiencies of Indian reckoning. Gradually, they incorporated the Indian zero into the notation of their own sophisticated system of mathematics and philosophy. Their great mathematician Al-Kharizmi (in whose honour we use the term algorithm) writes of the Indian calculating techniques that,[41]

> "When [after subtraction] nothing is left over, they write the little circle, so that the place does not remain empty. The little circle has to occupy the position, because otherwise there will be fewer places, so that the second might be mistaken for the first."

The Arabs did not originate a system of numerals of their own. Even in works of mathematics, they wrote out numbers word by word and accompanied them with parallel calculations in other systems, for example in Greek.[42]

Baghdad was a great cultural centre after its foundation in the eighth century and many mathematical works from India and Greece were translated there. In AD 773 the Caliph of Baghdad received a copy of a 150-year-old Indian astronomical manual, *Brahmasphutasiddhanta* (the 'Improved Astronomical Textbook of Brahma'), which used Indian numerals and place-value notation with a zero. Al-Kharizmi wrote his classic work on arithmetic forty-seven years later, explaining the new notation and its expediency in calculation. He introduced the practice of grouping numerals in threes, separated by commas, when writing large numbers that we still use today – as in 1,456,386 – unless the numbers are dates – year 2000, not 2,000. His book was translated into Latin and widely known in Europe from the twelfth century onwards.

The use of words or Greek alphabetical forms for numbers persisted until the tenth century when we see two sets of numerals develop, the 'East' and 'West' Arab numerals. An interesting feature of both these systems was their adoption of the Indian symbols for the numerals 1, . . . ,9 but not the zero. Instead, they developed a simple form of place-value notation that sidestepped it. If a numeral was denoting the number of tens then a dot was placed above it (e.g. 5 with one dot above meant 50), if it denoted the number of hundreds then two dots were placed above it, and so on. Thus the number three hundred and twenty-four, which we write as 324, would have appeared as

$$3\overset{..}{2}4$$

but 320 would have been written as $3\overset{..}{2}$ and 302 as $\overset{..}{3}2$. Later, the East Arabs introduced the small circle for the zero system and aligned their notation fully with the Indian convention.

The introduction and spread of the Indo-Arab system of numbers into Europe is traditionally credited to the influence of a Frenchman, Gerbert of Aurillac (945–1003). He became acquainted with Arab science and mathematics during long periods spent living in Spain and was extremely influential in directing theological education in France, and later in other parts of Europe. He had humble beginnings but received a good education in a monastery, and went on to hold a succession of High Church offices, as Abbot of Ravenna and Archbishop of Rheims, before ultimately being elected Pope Sylvester II in 999. Gerbert was the first European to use the Indo-Arab system outside Spain and was one of the most important mathematicians of his time, writing on geometry, astronomy and methods of calculation: a unique mathematical pope.

Gradually, the advantages of the Indo-Arab system became compelling and by the thirteenth century it was quite widely used for trade and commerce. Yet, despite its efficiency, there was opposition. In 1299, a law was passed in Florence forbidding its use. The reason was fear of fraud. Its rival, the system of Roman numerals, is not a place-value system and contains no zero. In the days before the invention of printing, all financial records were handwritten and special measures had to be taken to prevent numbers being illicitly altered by unscrupulous traders. When a Roman number I appeared on the end of a number, for example in II, denoting 'two', it would be written IJ to signal that the right-hand symbol was the end of the number. This prevents it being turned into III (but, alas, not into XIII) and is akin to our practice of writing 'only' after the amount to pay on a personal cheque. Unfortunately, the Indo-Arab system appeared wide open to fraud of this sort. Unlike the Roman system the addition of a numeral on the end of any number creates another larger number (most such additions did not create a meaningful number in the Roman system). Worse still, the zero symbol lays itself open to artistic elaboration into a 6 or a 9. These problems played an important role in bolstering natural inertia and conservatism which held up the wholesale introduction of the Indo-Arab system amongst the majority of merchants in Northern Europe until well into the sixteenth century.[43]

THE EVOLUTION OF WORDS FOR ZERO

"It was said that all Cambridge scholars call the cipher aught and all Oxford scholars call it nought."

Edgworth[44]

We have seen that our numerical zero derives originally from the Hindu *sunya*, meaning void or emptiness, deriving from the Sanskrit name for the mark denoting emptiness, or *sunya-bindu*, meaning an empty dot. These developed between the sixth and eighth centuries. By the ninth century, the assimilation of Indian mathematics by the Arab world led to the literal translation of *sunya* into Arabic as *as-sifr*, which also means 'empty' or the 'absence of anything'. We still see a residue of this because it is the origin of the English word 'cipher'. Originally, it meant 'Nothing', or if used insultingly of a person it would mean that they were a nonentity – a nobody – as in *King Lear* where the Fool says to the King[45]

> "Now thou art an 0 without a figure. I am better than thou art now. I am a fool, thou art nothing."

The path to this meaning is intriguing. The Arab word *sifr* was first transcribed into medieval Latin in the thirteenth century in the two forms *cifra* or *zefirum*, and into Greek as τσιφρα, which led to their use of the letter tau, τ, as an abbreviation for zero. But the two Latin words acquired quite different meanings. The word *zefirum* (or *cefirum*, as Leonardo of Pisa[46] wrote it in the thirteenth century) kept its original meaning of zero. In fourteenth-century Italian, this second form changed to *zefiro*, *zefro* or *zevero*, which was eventually shortened in the Venetian dialect to *zero* which we still use in English and French. This same type of editing down was what reduced the currency from *libra* to *livra* to *lira*.

By contrast, the word *cifra* acquired a more general meaning: it was used to denote any of the ten numerals 0, 1, 2, . . . , 9. From it comes the French *chiffre* and the English *cipher*. In French, the same ambiguities of meaning exist as in English. Originally, *chiffre* meant zero, but like *cipher* came to mean any of the numerals. The merger of the ideas for zero and Nothing gave rise to the name 'null' being used either to denote 'Nothing' or the circular symbol for zero. This meant a 'figure of nothing', or *nulla figura* in Latin. John of Hollywood (1256) writes in his *Algorismus* of the tenth digit that provides the zero symbol:

> "The tenth is called *theca* or *circulus* or *figura nihili*, because it stands for 'nothing'. Yet when placed in its proper position, it gives value to the others."[47]

A fifteenth-century French book of arithmetic for traders tells us:

> "And of the ciphers [*chiffres*] there are but ten figures, of which nine are of value and the tenth is worth nothing [*rien*] but gives value to the others and is called zero [*zero*] or cipher [*chiffre*]."[48]

It is interesting that both these commentators write of ten symbols, including the zero. We can conceive of how finger-counting culture might have devised a system in which the ten fingers were used to denote the quantities 0 to 9 rather than 1 to 10. Yet the conceptual leap needed to associate that first finger with nothing would have been vast. Needless to say, no finger counters did that, but we don't know what use they made of a hand displaying no fingers to convey the intuitively simple piece of information that they had nothing left.[49]

In German, the ambiguity between the word for numbers and for zero was broken, with numbers called *Figuren* while the words *cifra* or *Ziffer* were used for zero.[50] The English word 'figures' was, as now, a synonym for numerals and 'being good with figures' became a familiar accolade for anyone possessing some ability as a computer.[51]

The terms *theca* and *circulus* ('little circle') are sometimes encountered as synonyms for zero. Both refer to the circular form of the sign for zero. *Theca* was the circular brand burned into the forehead or the cheeks of criminals in the Middle Ages.

A FINAL ACCOUNTING

> "A place is nothing: not even space, unless at its heart – a figure stands."

Paul Dirac[52]

So far, we have seen some of the history of how we inherited the mathematical zero sign that is now so familiar. It is part of the universal language of numbers. Obvious though it may now seem to us, very few ancient cultures appreciated its need and most of those that did needed a little prompting from its inventors. The system of attributing a different value to a numeral according to where it is located in a list was one of the greatest discoveries that humanity has ever made. Once made, it requires the invention of a symbol that signals that no value be attributed to an empty location in the list. The Babylonians and the Indian cultures first made these profound discoveries and their invention spread to Europe and beyond around the globe by the channel of the Arab culture and its sophisticated interest in mathematics, philosophy and science. Strangely, the ancient Greeks, despite their extraordinary intellectual achievements, failed to make these basic discoveries. Indeed, we have seen that their approach to the world and the use of logic to unravel its workings was a serious impediment to the genesis of the zero concept. They demanded a logical consistency of their concepts and could not countenance the idea of 'Nothing' as a something. They lacked the mystical thread that could interweave the zero concept into a practical accounting system. The fact that the Indian system worked, and transparently so, was sufficient to justify its spread. Their affinity to the philosophical concept of Nothing as a desirable thing in itself, not merely the absence of everything else, allowed the zero symbol to accrete a host of other meanings that persist today in our words for Nothing. Nothing started as a small step but it brought about a giant leap forward in the effectiveness of human thinking, recording and calculation. Its usefulness and effectiveness in commerce, navigation, engineering and science ensured that once grasped it was a symbol that would not be dropped. For, as Napoleon Bonaparte pointed out, 'The advancement and perfection of mathematics are ultimately connected with the prosperity of the state.'

Zero was like a genie. Once released it could not be restrained, let alone removed. Once words existed for the concept that the zero symbol represented, it was free to take on a life of its own, unconstrained by the strictures of mathematics, and even those of logic. The mathematicians had played a vital role in making legitimate the concept of Nothing in a place where it was easiest to define and control. In the centuries to follow, it would emerge elsewhere in different guises, with even deeper consequences, and more puzzling forms.

chapter two

Much Ado About Nothing

"Among the great things which are found among us the existence of Nothing is the greatest."

Leonardo da Vinci[1]

"Nothing really matters."

Queen

WELCOME TO THE HOTEL INFINITY

". . . the library contains . . . Everything: the minutely detailed history of the future, the archangels' autobiographies, the faithful catalogue of the Library, thousands and thousands of false catalogues, the demonstration of the fallacy of the true catalogue, the Gnostic gospel of Basilides, the commentary on that gospel, the commentary on the commentary on that gospel, the true story of your death, the translation of every book in all languages, the interpolations of every book in all books."

Umberto Eco[2]

"Nothin' ain't worth nothin', but it's free."

Kris Kristofferson & Fred Foster[3]

The development of European thinking about the puzzles created by Nothing is a story about grasping two horns of a dilemma. Five hundred years ago, if you were a philosopher you might have had to get a grip on the slippery abstract concept of Nothing and persuade your peers that Nothing could be something after all – not least, something worth studying. But if you were a practising scientist, a 'natural philosopher', you faced the deeper paradox of whether there could exist a physical Nothing: a perfect vacuum of empty space. Worst of all, both of them risked serious disapproval from the religious status quo for letting their thoughts stray into such potentially heretical territory. Nothing was an ultimate issue, what nowadays we might call a 'meaning-of-life question': a question whose answer has the potential to unsettle the foundations of entire edifices of thought, carefully arranged to withstand the perturbations of new ideas. Any theology that had doctrines about the beginning of the world, and from whence the world had sprung, had to have a view about Nothing. Nor is any answer quite as simple as it seems. Say 'Nothing at all' to the question of what was before the beginning of the world and trouble could be in store.

It does not immediately occur to us that Nothing might be an impossible state. But there was a time when it was hard for many to think otherwise. Plato's influential philosophy taught that the things that are seen around us are just imperfect manifestations of a collection of perfect ideal forms – blueprints from which all material things take their character. These forms are eternal, indestructible and invariant. Remove every material thing in the physical universe and the Platonist would still hold that these eternal forms exist. They are the 'mind of God' in modern parlance.[4] If we were to assume that Nothingness is one of these forms then it is impossible to conceive of an imperfect manifestation of it that would still merit the title Nothingness. A vacuum that contains a single thing is no sort of vacuum at all.

The problems facing anyone thinking about Nothing are not unlike those that face anyone contemplating what we call 'infinity'. They are problems because we stand firmly and finitely between the two extremes marked by zero and infinity. At first they appear intimately linked. Divide any number by zero and we get infinity. Divide any number by infinity and we get zero. But just like the ski resort full of girl-chasing husbands and husband-chasing girls, the situation is not as symmetrical as it might first appear. For a mathematician, the idea of zero is straightforward and uncontroversial: we see concrete examples of it when the quantity of any commodity is exactly exhausted. It obeys simple rules of addition and multiplication.[5] But infinity is quite another matter. Some currents of mathematical opinion have, in the past, argued that mathematics should only be allowed to deal with finite collections of things that can be enumerated in a step-by-step fashion. The more conventional view is that formal infinities are all right in mathematics but you must be very careful how you handle them. They do not obey the usual laws of arithmetic for finite quantities. Take an infinity away from an infinity and you can still be left with infinity: for example, the list of all whole numbers (1, 2, 3, 4, 5, . . .) contains an infinite number of odd numbers (1, 3, 5, 7, . . .) and an infinite number of even numbers[6] (2, 4, 6, 8, . . .). Take the infinite number of odd numbers away from the infinity of all numbers and you are left with an infinity of even numbers!

The problem of infinity is beautifully captured by the story of Hilbert's Hotel.[7] In a conventional hotel there are a finite number of guest rooms. If they are all taken then there is no way you can be accommodated at the hotel without evicting one of the existing guests from their room. But with an infinite hotel things are different. Suppose that one person turns up at the check-in counter of the Hotel Infinity with its *infinite* number of rooms (numbered 1, 2, 3, 4, . . . and so on, for ever), all of which are occupied. No problem: the manager asks the guest in room 1 to move to room 2, the guest in room 2 to move to room 3, and so on, for ever. This leaves room 1 vacant for you to take and everyone still has a room.

You are so pleased with this service that you return to the Hotel Infinity on the next occasion that you are in town, this time with an *infinite* number of friends. Again, this popular hotel is full. But again, the manager is unperturbed. He moves the guest in room 1 to room 2, the guest in room 2 to room 4, the guest in room 3 to room 6, and so on, for ever. This leaves all the odd-numbered rooms empty. There are an infinite number of them free to accommodate you and your infinitely numerous companions without difficulty. Needless to say room service was a little slow.

The contrast between zero and infinity is most marked when it comes to the physical realisation of these 'numbers'. Zeros are no problem – there are no wheels on my wagon – but no one knows whether infinities are physically manifested. Most scientists believe that they are not: their appearance in a calculation merely signals that the theory being employed has reached the limits of its validity and must be superseded by a new and improved version which should replace the mathematical infinity by a finite measurable quantity. In controllable situations, like the flow of a fluid, we can observe the physical situation in which the spurious infinity was predicted to occur, see that no physical infinity arises, and so be certain that more accurate mathematical modelling of the situation will exorcise the predicted infinity. However, there are more exotic situations, like that of the apparent beginning to the expansion of the Universe, where we can assure ourselves by observation that everything is physically finite. The situation being considered there is so singular in many respects that it is not clear why a physical infinity could not be present. Nevertheless, a large part of cosmologists' studies of this situation is directed towards trying to find a superior theory in which any beginning to the Universe is not accompanied by physical infinities.

Another contrast between zero and infinity is the psychological effect that each produces on human minds. In modern times there is little fear of zero – except when it appears too often in your bank balance – but many find the concept of the infinite to be awesome, mind-boggling, even terrifying, echoing Blaise Pascal's famous confession that 'The silence of infinite space terrifies me'. Nor are such sentiments confined to the seventeenth century. The famous Jewish philosopher Martin Buber, who died in 1965, wrote of how the mere thought of the infinite led him to contemplate suicide:

> "A necessity I could not imagine swept over me: I had to try again and again to imagine the edge of space, or its edge-

lessness, time with a beginning and an end or time without a beginning or end, and both were equally impossible, equally hopeless ... Under an irresistible compulsion I reeled from one to the other, at times so closely threatened with the danger of madness that I seriously thought of avoiding it by suicide."[8]

Existentialist philosophers have struggled to extract some sense from the contrast between Being and non-Being from a vantage point that sees all existence as deriving from human existence. The most well-known work of this sort is Jean-Paul Sartre's book *Being and Nothingness*, which contains tortuous ruminations over the meaning and significance of Nothingness. Here are some typical extracts:

> "*Nothingness haunts being*. That means that being has no need of nothingness in order to be conceived and that we can examine the idea of it exhaustively without finding there the least trace of nothingness. But on the other hand, nothingness, *which is not*, can have only a borrowed existence, and it gets its being from being. Its nothingness of being is encountered only within the limits of being, and the total disappearance of being would not be the advent of the reign of non-being, but on the contrary the concomitant disappearance of nothingness. *Non-being exists only on the surface of being*." [9]

Here, Sartre is contesting the idea, argued by Hegel, that Being and Nothingness are merely equal and opposite. He does not believe they can logically be contemporaries at all. Nor are they merely both 'empty abstractions, and the one is as empty as the other' as Hegel claimed, for the key feature that creates the asymmetry between them 'is that emptiness is emptiness of something'.[10] They are quite different.

GREEKS, BEARING GIFTS

"'I see nobody on the road,' said Alice.

'I only wish *I* had such eyes,' the King remarked in a fretful tone, 'to be able to see Nobody! And at that distance too! Why, it's as much as *I* can do to see real people, by this light!'"

Lewis Carroll

Ever since the early Greeks grappled with these problems the contemplation of Nothing has been bedevilled by paradoxes like those that afflict the contemplation of the infinite. Philosophers like Parmenides and Zeno marshalled these paradoxes to attack the self-consistency of the concepts of Nothing and infinity.

For Parmenides the Universe must be a unity. It is limited but fills all of space. Symmetry demands that it must be spherical in shape. A vacuum is impossible because it constitutes non-Being and contradicts the assumption that the Universe fills all space. Parmenides went so far as to protect his Universe from any intercourse with a vacuum anywhere else. He argued that things could not appear from Nothing or disappear into Nothing; he asked why such a creation from Nothing should have occurred at a particular moment and not sooner. Later supporters of the idea of creation out of Nothing, like Simplicius, answered this charge by suggesting that there might exist an orderly sequence of events, with individual forms of matter appearing one after the other. By reference to this logical sequence we can date any particular appearance.

European Christianity tried to wed together two pictures of Divine activity. One was the Greek picture of God as an architect who fashions the world out of pre-existing eternal material. The other was the Jewish tradition of God as the Creator of the World and all its properties out of Nothing. The Greek tradition held on to the belief that there was always something there originally from which the World was moulded. In this way it avoided having to wrestle with the concept of nothingness and thus with all the philosophical problems it carried with it. Greek philosophers recoiled from the concept of emptiness. The word chaos originally meant Nothing and shows us the anarchy that was attached to the very idea of regarding Nothing as something that had Being.

Philosophers like Parmenides and his disciple Zeno tried to defend their belief in the static unchanging nature of Being by a variety of ingenious arguments. Zeno's paradoxes of motion are amongst the gems of Greek thought and they were never refuted by other Greek thinkers, merely ignored. The Greek tradition focuses upon elements that do not change: points, lines, circles, curves and angles in geometry; numbers, ratios, sums and products in arithmetic. It is nervous of dealing with the limitless, and the opposition of zero and infinity attached a label saying 'beware' of both. Each dangled at the crumbling edge of thought. Aristotle saw them both as loose cannons in the logical structure of cause and effect. Nothing had no cause and no effects; no reason and no end. This presented a real quandary if one wanted to fit all concepts into a single harmonious logical structure, because as Brian Rotman pungently remarks:

> "For Aristotle, engaged in classifying, ordering and analysing the world into its irreducible and final categories, objects, causes and attributes, the prospect of an unclassifiable emptiness, an attributeless hole in the natural fabric of being, isolated from cause and effect and detached from what was palpable to the senses, must have presented itself as a dangerous sickness, a God-denying madness that left him with an ineradicable *horror vacui*."[11]

Greek philosophy and psychology could find no room in their indivisible Universe of unchanging Being for the sort of gap that the reality of Nothing would require. And so it simply could not be. One could not make something of Nothing. Aristotle defined the void to be a place where no body could be. This step would have allowed him to take off in many different philosophical explorations, moving East to contemplate the notions of non-Being and nothingness so beloved of the Indian thinkers. Instead, he concluded that the void could not exist. Eternal things occupy every place. There can be no state of perfect emptiness, devoid of Being.

Despite this antipathy to Nothing one does occasionally find some of the paradoxical wordplay that was to overtake English writers in the seventeenth century. The most striking is the encounter between Ulysses and the Cyclops, Polyphemos, created by Homer in *The Odyssey*.[12] Ulysses sets about lowering the one-eyed monster's guard by providing him with an abundance of wine. When asked by the Cyclops for his name he replies 'my name is Noman;[13] this is what my father and mother have always called me'. But the Cyclops vows to devour him, so Ulysses seizes his opportunity to blind the Cyclops with a burning stake from the fire. The Cyclops screams out to his neighbours for help: 'Noman is killing me by fraud! Noman is killing me by force!' No help comes, merely the replies that 'if no man is attacking you, you must be ill; when Jove makes people ill, there is no help for it'. Ulysses and his men slip by the blinded Cyclops, disguised by the fleeces of sheep, and make good their escape, but as they sail into the distance the Cyclops curse them never to return to their homes alive.

It is strange that this ancient epic bestseller did not stimulate any other Greek philosophers to take up the paradoxes of Nothing. They were ripe for the treatment that Zeno administered to the idea of infinity in memorable scenarios like those summarized in Figure 2.1.

Greek philosophy denied the concept of Nothingness right from its outset in the fifth and sixth centuries BC. Thales and his school in Miletus maintained first that 'something' can never emanate from Nothing or disappear into Nothing. He used this intuition to deny the possibility that the Universe could have appeared out of Nothing, a difficult idea to grasp and one that we in the Christian West have become comfortable with only because of two millennia of religious tradition. Parmenides was the first of the Greek philosophers to take the idea of 'non-Being' seriously and grapple with it in order to make sense of it. Thales had focused upon the attributes of Being and simply ignored the concept of non-Being. Parmenides maintained that non-Being did not exist but his exploration of

---

ZENO'S PARADOXES OF MOTION

The Race Course

There can be no motion because anything that moves must reach the half-way point of its journey before it reaches the end. So, to cover one metre of the race course, you must first cover half a metre, before that one quarter of a metre, before that one eighth, and so on forever. How is it possible to reach an infinite number of positions in a finite time?

Achilles and the Turtle

Achilles can run 400 metres in a minute while a turtle can run 40 metres a minute. The turtle starts 400 metres ahead of Achilles. Achilles can never overtake the turtle because after Achilles has run 400 metres the turtle is still 40 metres ahead of him. By the time Achilles has covered these 40 metres (in a tenth of a minute), the turtle is still 4 metres ahead of him, and so on, forever.

Figure 2.1 *Zeno's paradoxes of motion.*

these ideas never considered the practical questions of empty space and regions devoid of matter: of actually looking for a space that might potentially be empty. That more detailed step of speculative natural philosophy was taken by the Sicilian Empedocles, who later in life was to come to a grisly end by leaping into the active volcano on Mount Etna, perhaps ultimately coming to believe his delusions of divinity.

Empedocles imagined matter to contain pores of a mysterious light medium, called 'ether'. This quintessential part of the world was devised in order to avoid having to introduce the concept of empty space when trying to account for the granular structure of many forms of matter. In places where there was no evidence of any matter at all, Empedocles could maintain that there was always some of this ethereal substance, lighter than all known materials (except possibly air), permeating tiny pores and guarding

us against the horror of a perfect vacuum ever forming. To his credit, he
was not content to let the ether be simply a spoiler for the vacuum; he
envisaged emanations proceeding from the pores within bodies so that they
could influence one another in different ways. In some respects this intuition has a rather modern ring to it. Empedocles does not have the idea
(that Newton used about two thousand years later) that forces act instantaneously between different bodies. Rather, when a magnet pulls a piece of
iron towards it, the attraction takes a finite time to occur:

> "Why does a magnet attract iron? Empedocles says that the
> iron is drawn to the magnet, because both give off emana-
> tions and because the size of the pores in the magnet corre-
> sponds to the emanations of the iron ... Thus, whenever
> the emanations of the iron approach the pores of the mag-
> net and fit them in shape, the iron is drawn after the emana-
> tions and is attracted."[14]

This was the beginning of a belief in an ether. We shall see that it was
maintained in different forms until the start of the twentieth century. Its
original purpose was simply to avoid having to admit the existence of
empty space in the physical universe and to reconcile the picture of physical space and matter with the philosophical conceptions of Being and the
inconceivability of non-Being.

Empedocles was not just a philosopher. He made an important
experimental discovery in the course of his researches into human breath-
ing and the nature of air. He explains what he has observed about the
behaviour of a perforated vessel used to catch water,[15] displayed in Figure
2.2. He notices that if the water-catcher is submerged before the air has
been expelled from it then the water cannot flow into it,[16]

> "As when a girl, playing with the water-catcher of shining
> brass – when, having placed the mouth of the pipe on her
> well-shaped hand she dips the vessel into the yielding
> substance of silvery water, still the volume of air pressing

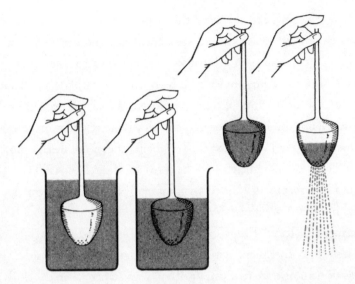

Figure 2.2 *The ancient water-catcher experiment. Immerse the perforated vessel and then seal the tube with a finger. On removing it from the water the water remains trapped in the vessel but when the finger is removed from the tube it escapes. An explanation for this behaviour was a challenge to scientists and philosophers for more than 2000 years.*

> from inside on the many holes keeps out the water, until she uncovers the condensed stream [of air]. Then at once when the air flows out, the water flows in in an equal quantity."

He is on the verge of deducing something about the pressure exerted by the air in the Earth's atmosphere. Two thousand years would pass before Torricelli provided the correct explanation for the behaviour of devices like this.

Anaxagoras, like Empedocles, lived in the middle of the fifth century BC, working first in Ionia and then in Athens. Like Empedocles, he also denied the existence of empty space and believed strongly in the conservation of the 'essence' of the world. This conservation principle meant that things could not appear out of Nothing, or disappear into it. It is an idea that is similar in spirit to our modern concept of the conservation of energy. Anaxagoras viewed 'creation' as the bringing of order into a state of primordial chaos rather than as an event from which the World came into being out of Nothing. He also used this conservation principle to understand how things change from one substance into another; for example, how fruit or other forms of food that we eat can turn into flesh and bones. He believed that something must be passed on in each of these changes, that there are 'seeds' within all forms of matter which are passed on but neither created nor destroyed. 'For in everything there is a portion of everything.' One might even view these seeds as being the molecules of modern chemistry. Yet these ingredients were held to be infinitely divisible, so that space could be continuously filled with matter. No need for Empedocles' pores, and no room for empty space either.

Anaxagoras shared Empedocles' fascination with the water-catcher and repeated that experiment, extending it by compressing air inside wine-skins so as to demonstrate that the air offers a resistance when the skins are stretched. From this, he concluded that air is not the same thing as empty space and that we have no observational evidence for the existence of empty space. A subtle thinker, he was the first philosopher to recognise that our observations of the world are conditioned by the frailty of our senses. Our ability to decide whether one thing is really different from another (his favourite example was distinguishing very similar shades of colour) is just a reflection of our senses and because of 'the weakness of the sense-perceptions, we cannot judge truth'. Our senses are sampling partial information about a deeper reality that they cannot fully apprehend. His ideas were ones that would be used by the Greek atomists who came after him as a fundamental feature of their picture of the world.

The atomists maintained that all matter was composed of atoms, tiny indivisible particles (the Greek word *atomos* means having no parts), which were eternal, indivisible and unchangeable. Atoms moved through empty space and their different degree of clustering from place to place was responsible for changes in density and the distinctive properties of different forms of matter. This powerful picture of the world was appealing because of its simplicity and wide applicability. It was proposed first by Leucippus of Miletus in the mid-fifth century BC, developed further by his student Democritus, and eventually upgraded into an entire philosophical system by Epicurus of Samos (341–270 BC), after which it became extensively known. Even so, today its most memorable articulation is to be found in the remarkable poem *De Rerum Natura* (*On the Nature of Things*) composed by the Roman poet Lucretius in honour of Epicurean atomism in about 60 BC.

Leucippus has the double distinction of introducing the concept of matter being composed of identical basic units and of taking seriously the idea that there does indeed exist something called empty space in which these atoms move. Here we see for the first time the concept of a true vacuum being rigorously employed as an axiomatic part of a natural philosophy. Because the world was differentiated into atoms and the void in which they moved, the vacuum was necessary for any movement or change to be possible, and Leucippus reminds us that[17]

> "unless there is a void with a separate being of its own, 'what is' cannot be moved – nor again can it be 'many', since there is nothing to keep things apart."

Atoms could differ in concentration, in shape and in position, but they could not appear and disappear from or into Nothing. This immutability of atoms rules out any possibility that they contain regions of vacuum. They must be solid and finite in size. It may be significant that Leucippus spent some time as a pupil in the philosophical school of Elea where Zeno had worked, and where his paradoxes of the infinite were much studied. Zeno had demonstrated some of the bizarre paradoxes that could occur if you considered a process of halving things indefinitely. For example, one of his paradoxes of motion invites us to contemplate how it is possible to walk, say, to the door of our room one metre away. First, we must cross half a metre, then half of half a metre, then half of half of half a metre, and so on, *ad infinitum*. It appears that we will never be able to reach the door because we have to cover an infinite number of distances! It is possible that these awkward problems of dealing with things that were allowed to become arbitrarily small convinced Leucippus of the importance of having a smallest possible size for his atomic units of matter to avoid such paradoxes. There were physical reasons as well. Epicurus argued that allowing matter to be infinitely divisible would result in the irreversible destruction of its identity, slipping ultimately into non-existence, or give rise to aggregates of matter that were too fragile to persist. This was a far-reaching step because it drew a sharp distinction between mathematical and physical reality: in the former, infinite division of any quantity was possible; in the latter, it was not. You had to choose which mathematical structure to apply to physical existence.

According to Epicurus,[18] atoms must also have a maximum possible size in order to explain why they are not seen with the naked eye. Democritus is silent[19] on this point, but he agrees with all the other atomists that the number of atoms in the Universe, like its size and age, is infinite. Thus, their conception of the Universe is as a vacuum of infinite size filled with moving, solid, indivisible particles of different shapes and sizes.[20] Lucretius poetically describes how it can be that the random motion of these imperceptible atoms can give rise to everyday objects that seem to be steady and unchanging:[21]

> "Although all the atoms are in motion, their totality appears to stand totally motionless . . . This is because the atoms all lie far below the range of our senses. Since they are themselves invisible, their movements also must elude observation. Indeed, even visible objects, when set at a distance, often disguise their movements. Often on a hillside fleecy sheep, as they crop their lush pasture, creep slowly onward, lured this way or that by grass that sparkles with fresh dew, while the full-fed lambs gaily frisk and butt. And yet, when we gaze from a distance, we see only a blur – a white patch stationary on the green hillside."

There is a curious parallel between the atomists' picture of atoms separated by the void and Pythagoras' picture of numbers. Pythagoras and his followers believed that everything could be expressed by numbers and these numbers possessed intrinsic meanings, they were not merely ways of expressing relationship between things. If two quite different things possessed an element of threeness, or fiveness, then they were deeply related by a fundamental harmony. Like the atomists, the Pythagoreans required a void to exist in order to maintain the identities of things. For the atomists, it was empty space that separated atoms and allowed them to move. For the Pythagoreans, everything was number: the void existed between numbers. Aristotle reports that the Pythagoreans maintained that[22]

> "the void exists . . . It is the void which keeps things distinct, being a kind of separation and division of things. This is true first and foremost of numbers; for the void keeps them distinct."

The atomists were not the only ancient philosophers to have strong views about the vacuum. From the third century BC, there emerged a completely different theory of the nature of things. It became known as Stoicism, after its first adherents were dubbed Stoics because they chose to meet under a painted corridor (*stoa*) on the north side of the market place in Athens. Its founders were Zeno of Cition (not to be confused with Zeno of the paradoxes), Chrysippus of Soli in Cilicia, and Poseidonius of Apamea in Syria.

In complete contrast to the atomists' dogma, the Stoics believed that all things were a continuum, bound together by a spirit – an elastic mixture of fire and air – or *pneuma*, that permeated everything. No empty space could exist within or between the component pieces of the world, but this did not mean that there couldn't exist any empty space at all. Quite the contrary, the Stoics' Universe was a finite continuous island of material diffused by pneuma, but sitting in an infinite empty space.[23] The void was the great beyond and the pneuma bound the constituents of the world together so as to prevent them diffusing out into the formless void.

The Stoic conception is of interest to us now because the pneuma was a forerunner of the long-lasting idea that space is filled with a ubiquitous fluid, an *ether*, which can be acted upon and which responds to the actions of other material. The Stoics envisaged their ether as a medium through which the effects of sound or other forces could propagate, just as when we disturb the surface of water in one place we can see the waves emanating outwards over the surface to create effects elsewhere, causing a nearby floating leaf to oscillate up and down.

Remarkably, neither the views of the atomists nor those of the Stoics proved influential over the next fifteen hundred years. The dominant picture of the natural world that emerged from Greek civilisation and wedded itself to the Judaeo-Christian world view was that of Aristotle. Aristotle's approach to natural phenomena was dominated by a search for purpose in motion and change. While this teleological perspective could be of help in understanding what was going on in the natural world, or in the study of human psychology, it was a real obstacle to the study of problems of physics and astronomy. Aristotle's picture of Nature was extremely influential and his views about the vacuum fashioned the consensus view about it until the Renaissance.[24] He rejected the possibility that a vacuum could exist, either in the world as the atomists maintained, or beyond it as the Stoics believed. The Aristotelian universe was finite in volume; it contained everything that exists; it was a continuum filled with matter; space was defined by the bodies it contained. But unlike the dynamical ether suggested by the Stoics, Aristotle's continuous ether was static and passive, eternally at rest.

ISLAMIC ART

"Humility collects the soul into a single point by the power of silence. A truly humble man has no desire to be known or admired by others, but wishes to form himself into himself, to become nothing, as if he had never been born. When he is completely hidden to himself in himself, he is completely with God."

Isaac of Ninevah (AD 600)[25]

When compared with ancient Greek or later Western representational art, the intricate mosaics and tessellations of Islamic art seem like an ancient form of mathematical art: computer art before there were computers. We can picture their teleported ancient creators manipulating fractals and modern tiling patterns to continue a tradition that vetoed the representation of living things. Their patterns are extremely revealing of their religious views. God alone was infinite. God alone was perfect. But by creating finite parts of patterns that were evidently infinite it was possible to capture a little piece of the Divine in a humble yet inspiring manner. The partial character of the design served to reinforce the frailty and finiteness of humanity in contrast to God's infinity.

Islamic art directed the mind towards the infinite by creating regular patterns that could be infinitely repeated. These designs have become familiar to us through the work of the Dutch artist Maurits Escher and the mathematical designers he inspired. Born in Leeuwarden, Holland, in 1898, Escher began his artistic career as a landscape artist, painting little Mediterranean towns and villages. But his life's work was changed in the summer of 1936 by a visit to see the fabulous designs of the Alhambra, in Granada, Spain (Figure 2.3).

Escher was deeply impressed by the intricate patterns he saw and the fabulous geometric precision of the creators of this fourteenth-century Moorish palace. He spent many days studying the detailed patterns and periodicities, and went away to develop his own synthesis of symmetry and impossibility. Unlike the patterns of the Alhambra, Escher animated his designs with living creatures: fish, birds, winged horses and people. This expression of the abstract by means of recognisable images was, he remarked, the reason for his 'never-ceasing interest' in these patterns.

In Islamic art we see how the Moslems celebrated infinity where the Greeks feared it. They made it the hidden engine of their artistic creations. While not quite on central stage, it was never far away in the wings. The treatment of zero and Nothingness is just as confident. Rather than sweep Nothing away under the carpet as a philosophical embarrassment, the Islamic artists simply saw the void as a challenging emptiness to be filled. No blank space could be left alone. They filled friezes and surfaces with intricate patterns.[26] This urge seems to be shared by human cultures the world over. Wherever anthropologists look they find elaborate decorations.

Figure 2.3 *Islamic decorations from (top) Badra in Azerbaijan, and (bottom) the Alhambra Palace in Granada, Spain.*

We just do not like an empty space. As we saw with the Mayan need to fill their mathematical pictograms with an image for Nothing, the human mind longs for pattern and for something to fill any void. The great art historian Ernst Gombrich termed this impulse to decorate the *horror vacui.* It inspires a wealth of persistent procedures, sometimes linking different parts, sometimes filling in space, allowing a network of growing intricacy to emerge and develop.

ST AUGUSTINE

"Miracles are explainable; it is the explanations that are miraculous."

Tim Robinson

In medieval and Renaissance thought the paradoxical aspects of the something that is Nothing became interwoven with the doctrines and traditions of Christian theology. These doctrines were founded upon the Jewish tradition of turning away from Nothing because it was the antithesis of God. God's defining act was to create the world out of Nothing. What stronger evidence could there be that Nothing was something undesirable: a state without God, a state which He had acted to do away with. Nothing was the state of oblivion to which the opponents and enemies of God were dispatched. Any desire to produce a state of nothingness or empty space was tantamount to attempting what only God could do, or to remove oneself from God's domain. A single Divine creation of everything out of Nothing was a basic tenet of faith. To speak seriously of the void or of empty space was atheistic. It countenanced parts of the Universe where God was not present.

The most innovative thinker to grapple with the problem of synthesising the Greek horror of the void with the Christian doctrine of creation out of Nothing was Augustine of Hippo (354–430), pictured in Figure 2.4. He had a broader and deeper view of what creation should mean. It needed to be something more than the mere refashioning of primitive pre-existent materi-

Figure 2.4 *St Augustine.*

als into an ordered cosmos, something more than the unfurling of the cosmic scene at some moment in the distant past. Rather, it must provide the
ground for the continued existence of the world and an explanation for time
and space itself. It was for him a total bringing into being. Nothingness was
therefore an immediate precursor state to the one that God sustains. This
makes it more negative in its attributes than merely not being what is now in
the Universe. It was characteristic of being apart from God.

Augustine equated Nothing with the Devil: it represented complete
separation from God, loss and deprivation from all that was a part of God,
an ultimate state of sin, the very antithesis of a state of grace and the pres-

ence of God. Nothing represented the greatest evil. This was the 'something' that he believed non-Being to be. This formula led Augustine into dangerous waters because by introducing Nothing into the realm of Being he admitted that there was something that God lacked before he created the world. This difficulty he sidestepped, along with other problems about the beginning of time, by arguing that when God created the world he created time as well. There was no 'before' the first moment of time and so no time when God needed to change an unsatisfactory state of affairs.

These pieces of theological legerdemain were never entirely persuasive and centuries later they led Thomas Aquinas to create a fuller negative theology in which the attributes of God were only to be spoken of negatively: He was not finite, not temporal, unchangeable, and so forth. Aquinas supported the Aristotelian abhorrence of Nothing by viewing the creation of the world as an annihilation of Nothing in an act of Divine creative transformation. Yet, despite this careful circumscription, the Church was wary of Nothing and its mathematical representations during the tenth to thirteenth centuries. It tried to keep Nothing confined to the realm of arithmetic symbols where zero could be relegated to a harmless place holder on a counting board, far from the philosophical implications that the Indians had embraced but from which the Greek-Christian synthesis had recoiled.

There were two threads to the theological writings: one which drew out the nature of the nothingness from which creation had sprung; another which emphasised the nothingness and ephemerality of all temporal things. Both were directed at refuting the dualist heresy that the world was created out of pre-existing matter, rather than out of Nothing. The first thread was the preserve of serious theological treatises whereas the second was the substance of metaphysical poets trying to prove the nothingness of life when viewed in the cosmic scheme of things.[27]

It is important to recognise that although Christian doctrine included the notion of creation out of nothing (*creatio ex nihilo*), it did not include the idea that the creation was *caused* by nothing. The cause[28] of the creation is God, not some latent property of the void. God always exists but the Universe just lacks a material cause to initiate its structure. Aquinas argued

that if there is absolutely nothing – no Universe, no God, no Being at all –
then nothing can appear. For to cause itself, a being would have to exist in
order to give itself existence and this, he claims, is absurd. So if absolutely
nothing ever existed in the past, nothing exists now.[29]

THE MEDIEVAL LABYRINTH

"But if there is a void above and a void below, a void within
and a void without, he who is intent on escaping void has
need of a certain imaginative mobility."

Robert M. Adams[30]

It is easy to skip over the Middle Ages as though they were a time of darkness and delusion, an antechamber in the history of scientific ideas that
awaits the arrival of Copernicus, Galileo and Newton. But in order to
understand why scientific ideas about space and the vacuum developed in
the way they did, when they did, it is important to take some snapshots of
the way in which human thinking about the concept of Nothing developed
from the time when Aristotle's ideas held sway until the early eighteenth-
century arguments between Newton and Leibniz. Scholars of all complex-
ions struggled for more than five hundred years to harmonise subjects like
the nature of space, infinity and the vacuum. Their task was made more
difficult by their need to relate all these concepts to the nature and capacity
of God. The synthesis of Aristotle's philosophy and Christianity created
a complex web of philosophical ideas whose theological consistency was
more important than the mere assimilation of experimental facts; not because those facts were regarded as of little relevance, but because their significance was often ambiguous and they could be incorporated into the
World model in a variety of ways consistent with their overall World view.

As a result of Aristotle's rejection of the idea that a separate vacuum
could exist, on the grounds that it was logically incoherent,[31] it was believed almost universally in the early Middle Ages that Nature abhorred

the creation or persistence of any vacuum state. Almost all scholars believed that it was not possible to create a vacuum within the universe of space that we experience and see around us, a so-called *intracosmic void*. Things became more complicated when attention was turned to consider the possibility that there might exist an infinite *extracosmic void* beyond our finite, spherical, Aristotelian universe. This idea began to have credence in the fourteenth century and became very widely accepted over the next three hundred years.

Medieval philosophers inherited a strong Aristotelian opposition to the vacuum. In order to leave no hiding place from his arguments for its non-existence, Aristotle defined the vacuum carefully. He characterised it as that in which the presence of a body is possible, although not actual. Aristotle attempted to show that admitting the idea of a vacuum would paralyse the Universe. Motion would be impossible because there was no reason to move one way or the other in a vacuum because it was necessarily the same everywhere and in every direction. There was neither 'up' nor 'down' and so no way for things to adopt their 'natural' motion. In any case, if motion did occur it would continue for ever because there would be no medium offering any resistance to its motion.[32] Perpetual motion was a reductio ad absurdum. Nor would it make sense for a moving body to stop anywhere in this perfectly homogeneous void: for why should it stop at one place rather than another?

Considerable attention was devoted by scholars in the thirteenth and fourteenth centuries to the idea that Nature disliked the presence of a vacuum so acted always so as to remove it or resist its creation. As ever, there were shades of opinion. Some – strict Aristotelians – maintained that it was impossible to make a perfect vacuum for even a fleeting instant of time. Others were content to permit the ephemeral existence of a vacuum so long as events inevitably overcame it and refilled it quickly with air or other material. They did not believe in the existence of a stable vacuum.

Some scholars like Roger Bacon were unhappy with a law of Nature that was negative. A rule like 'no vacuums allowed' could not be primary; it needed to be a consequence of a deeper positive principle about what

Nature did. Negative principles were very powerful vetoes but they allowed too many things to happen that were not seen. As a specific example, the workings of Empedocles' water-catcher, or clepsydra, were much debated. Bacon argued that the veto on the formation of a vacuum was insufficient to explain what is seen. The formation of a vacuum inside its walls could equally well be avoided by the walls imploding. Why does Nature choose to hang on to the water rather than cave in the walls of the vessel? What principle decides?

Another favourite puzzle that taxed medieval scholars was a simple example noticed first by Lucretius.[33] It is the problem of separating two smooth surfaces; for example, two flat sheets of glass or metal, as shown in Figure 2.5. The concern was that if they begin in perfect contact but are then suddenly pulled apart then a vacuum must be briefly formed when they separate: there must be a change from a state in which there is nothing between the sheets and one in which there is air between them. Here is Lucretius' ancient version of the problem:[34]

> "If two bodies suddenly spring apart from contact on a broad surface, all the intervening space must be void until it is occupied by air. However quickly the air rushes in all round, the entire space cannot be filled instantaneously. The air must occupy one spot after another until it has taken possession of the whole space."

The way this problem was investigated gives a wonderful insight into the ingenuity and serious medieval interest in these problems.[35] The theological stakes were surprisingly high, as we shall see.

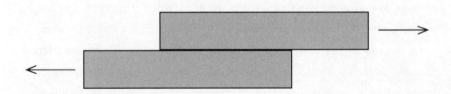

Figure 2.5 *Two parallel sheets sliding across their surface of contact.*

The Scholastics tried hard to show that this ancient puzzle, rediscovered by Bacon and others, did not allow even a short-lived vacuum to arise. Some claimed that while a vacuum could indeed form in principle if the surfaces were slid across each other so that they remained perfectly parallel, this could never happen in practice. A slight angle would develop between the surfaces and air would enter and gradually fill the gap between the surfaces bit by bit. Bacon turned the discussion around and argued that two smooth plane parallel surfaces could not be separated if they were in perfect contact (which would have a ring of truth about it for those who had tried it) unless they were first inclined. This was Nature displaying her resistance to the creation of a vacuum.

One of Bacon's English contemporaries, Walter Burley, saw through this suggestion, pointing out that the inclining of the surfaces makes no difference of principle at all. The ephemeral vacuum must still form, it merely lasts for a shorter time than if the plates are parallel when separated. Probing deeper still, he pointed out that perfectly parallel surfaces don't exist; but no matter, even though real surfaces always exhibit microscopic undulations which restrict their points of contact to occasional protrusions, all that the argument for an ephemeral vacuum needs is for there to be a single point of contact. When that point of contact is broken a vacuum must arise momentarily.

A notable opponent of the direction of this reasoning was Blasius of Parma, a student of motion and fluid flow. He argued that it *was* possible for plates to be separated by parallel motion yet not form a vacuum. This requires the particles of air to move at just the right speeds at just the right times to fill the void space immediately it forms.[36] But then he resorts to a very interesting argument, reminiscent of Zeno, that the vacuum can never form because there is no first moment when it exists — if you think there is, then halve it, and so on, ad infinitum. By denying the logical possibility of a first instant of separation for the plates, Blasius tried to eliminate the possibility of an instantaneous separation of the surfaces which would have allowed a vacuum to make even the briefest appearance.

Despite the ingenuity of these suggestions, the most widely accepted resolution of the dilemma stayed close to Aristotle's own treatment of a very similar problem. Aristotle had argued that there would always be some air trapped between two surfaces in contact, just as two surfaces which touch under water always get wet at their interface. Most people had regarded this as a conclusive and simple resolution of the surface contact problem until Bacon raised a simple objection. Forget about the two solid surfaces, he said. They are just a device to allow some air to be trapped in between the surfaces and avoid a vacuum forming. Suppose, instead, that you have only one surface and consider its interface with the water around it. Nothing lies between the surface and the water and so whenever they are separated a momentary vacuum must form![37]

Burley responded by claiming that there was still a thin film of air at the interface between a liquid and a solid. When you began to separate them it would quickly expand to fill up any potential void space. But what if the air was as rarefied as it could be and there was no scope for further expansion? Burley responded again by claiming that the surfaces would be inseparable in this situation. The only way of parting them would be to bend one of them and so create the inclined-surface problem that he and Bacon both believed to be the only way to effect a separation without producing a vacuum.

Despite his appeal to a specific physical process to avoid the vacuum's appearance, Burley felt he needed more protection from uncontrollable nat- ural events than this. What if a heavy rock should fall to the ground and expel all the air in between its surface and the ground at the point where they first made contact? Would there not be an instantaneous vacuum there? To stop all the air being expelled, he appeals to a celestial agent[38] which prevents the air yielding 'to the stone because [the stone] is held back by the superior agent, which powerfully seeks to prevent a vacuum'. If natural processes were inadequate to overcome the threat of a real vacuum forming then one needed to fall back on the cosmic censorial power of this supernatural agency to stave off the creation of Nothing from something. In the much- debated example of the water-catcher, the celestial agent could be invoked to explain why the water behaved 'unnaturally' by not falling downwards, rather than imploding the walls of the vessel to prevent a vacuum forming. This type of explanation, reminiscent of the *Just So Stories*, was not very persuasive.

Unfortunately, the celestial agent was annoyingly inconsistent in policing the formation of vacua. Others were quick to point out that on other occasions the universal agent did choose to stave off the vacuum by distorting the walls of the container, as occurs if the water freezes.

Alongside these detailed arguments about the ways in which Nature staves off the creation of an intracosmic void, there were centuries of debate about the existence of an *extracosmic void* – a vacuum beyond the physical Universe. Aristotle considered the idea briefly but rejected it along with the whole idea of the plurality of worlds. His definition of a vacuum as that in which 'the presence of a body is possible but not actual' demanded such a conclusion. For 'outside' the Universe there is no possibility of body and hence no vacuum. In this respect, as we have seen, he is diametrically opposed to the Stoic view that there exists an infinite extracosmic void.

The extracosmic void introduced further dilemmas for medieval philosophers. It was imagined to consist of 'imaginary' space; that is, space that could be imagined to exist even when bodies did not. We can imagine all sorts of things that seem to go on for ever, like the list of all numbers, and these may be imagined to 'live' in this infinite imaginary space.[39] Although it couldn't contain ordinary matter it had a property that proved crucial in the subsequent development of ideas. It was completely filled with the presence of God and was both the expression of God's immensity and the means by which His omnipresence was achieved and maintained. This considerably narrowed the options when it came to pinpointing its properties. Try to make the extracosmic void finite, or endow it with dimensions, and you risk reaching a heretical conclusion about the nature of God. For in order that God remains omnipresent yet indivisible, He needs to be wholly located at every single point of the infinite space of the extracosmic void: One who 'is an infinite sphere whose centre is everywhere and circumference nowhere'.[40]

The key moment in the early stages of these debates was the famous Paris condemnations of 1277 in which Bishop Etienne Tempier strove to reassert the doctrines of God's power to do whatever He chooses. Before

Tempier's intervention, there was a widespread belief amongst theologians that Aristotelian philosophy showed that God was constrained in various ways. For example, God could not make two and two make five; God could not create a plurality of worlds; and, inevitably, there was a veto that God could not cause a movement of things that would produce a local vacuum. By denying these restrictions on God's power, Bishop Tempier made space for an extracosmic vacuum. For if many worlds exist what lies between them? And if God should choose to move our whole world in a straight line then what would remain in its former place? 'A vacuum' was the answer that the prompter was whispering from the wings. And if you didn't hear, then beware of suggesting to the Bishop that God could not create a vacuum. After 1277, the vacuum became admissible because any attempt to exclude it on philosophical grounds was tantamount to limiting the power of God.

Another great medieval question was whether a vacuum had existed before the creation of the world. Aristotle had denied the possibility that the world (or anything else) could be created from Nothing. The original Aristotelian scenario of an eternal, uncreated Universe had the drawback of clashing with Christian doctrine and so the more appealing alternative was a version in which the world had been created from a pre-existing void containing nothing. Yet this was not entirely without problems of its own. It required the existence of something eternal that was seemingly independent of God. It was this stance that spawned infamous questions like 'What was God doing before the creation of the world?' and the development of Augustine's response that entities like time and space were created along with the Universe so there was no 'before'.

By the sixteenth century the tide had begun to turn. The rediscovery of lost texts by Lucretius and accounts of Hero's ancient experiments on pressure inspired assaults on Aristotelian dogma. The horror at allowing a vacuum to form abated and there was a change of attitude to the existence of an infinite void space which would alter the relationship of God to that space, culminating in a complete decoupling of the scientific and theological debates about the nature of space and the vacuum.

In the sixteenth and early seventeenth centuries those who began to subscribe to the Stoic cosmology, with its finite cosmos surrounded by an infinite extended void, were all agreed on many of the attributes of the surrounding void: it was the same everywhere, immutable, continuous and indivisible, and offered no resistance to movement. But what was new was a growing disagreement as to God's relationship to the infinite void space. Notable atomists like Pierre Gassendi denied that the infinite void had anything to do with the attributes of the Deity. A third way was provided by the philosopher Henry More who, while regarding space as an attribute of God, also regarded God as an infinite extended Being. More is interesting primarily because his views seem to have influenced Isaac Newton's views about space. Newton introduced God as a three-dimensional presence everywhere and as the underpinning intelligence behind the mathematical laws of Nature. Indeed, he introduced a new form of the ancient Design Argument for the existence of God, appealing to the fortuitous structure of the laws of Nature rather than their outcomes as evidence for a Grand Designer behind the scenes.[41] Newton clung to the Stoic picture of a finite world surrounded by an infinite void space. He could imagine an empty space but not the absence of space itself. Thus space was something that was quite independent of matter and motion. It was the cosmic arena in which matter could reside, move and gravitate. Newton writes that,[42]

> "we are not to consider the world as the body of God, or the several parts thereof as the parts of God. He is a uniform Being, void of organs, members or parts, . . . being everywhere present to the things themselves. And since space is divisible *in infinitum*, and matter is not necessarily in all places, it may also be allowed that God is able to create particles of matter of several sizes and figures, and in several proportions to space, and perhaps of different densities and forces, and thereby to vary the laws of Nature, and make worlds of several sorts in several parts of the Universe. At least I see nothing of contradiction in this."

For Newton, the extracosmic void space was entirely real and not in the least imaginary. When he was preparing the 1706 edition of his *Opticks* for the press he considered adding to his list of 'queries' – a series of far-reaching questions and speculations about the physical world – a final question asking[43]

"what the space that is empty of bodies is filled with."

These Newtonian views about the reality of the extracosmic void space and its relationship to God were articulated by his champion, Samuel Clarke, in a famous debate with Leibniz. Leibniz disagreed fundamentally with Newton. He denied that the infinite void even existed and objected to Newton's idea that we equate it with the immensity of God. He saw how difficult it was to sustain a relationship between God and space and opposed any such attempt. In the end, his view about the separation of God from space prevailed amongst philosophers and theologians, even though the infinite void space of Newton was retained by scientists.

Newton's God was no longer located in the void beyond the material world. The great idea of the Scholastics, that God was inextricably linked to the nature of space and to its infinite extent, lived long enough to influence Newton's great conception of the world and the laws of motion and gravity that governed it, but by the end of the eighteenth century the theological complexion of the problem of space had been eroded. The proposals for explaining God's omnipresence in space had lost credibility and played no further role in understanding the things that were seen. The Almighty could then be removed without reverberations spreading into the theological domain. Gradually, it was God's transcendence rather than His omnipresence that would become the centrepiece of the theologian's discussion of God. Once this transformation was complete, God needed no place in the infinite void of space that the astronomers took as the backdrop for the finite world of matter and motion. It was an arena that finally allowed mathematical deductions to be made without the need for a theological conscience. The vacuum was at last safe for scientists to explore.

WRITERS AND READERS

"Now is the discount of our winter tents."

Advertisement in Stratford-upon-Avon camping shop[44]

Not everyone spoke so seriously. In order to sidestep the risk of being accused of blasphemously toying with the demonic concept of empty space, writers and philosophers cloaked their thoughts in more playful deliberations, inventing and pursuing paradoxes and puns in a way that could always be defended as undermining the coherence of the concept of empty space regardless of the true intent. The paradox that would bring the argument to an end could always be defended as a reductio ad absurdum. The American commentator Rosalie Colie concludes her study of the poems and paradoxes of Nothing that were all the rage in the fifteenth and sixteenth centuries with the opinion that the writers of these paradoxes

> "were engaged in an operation at once imitative and blasphemous, at once sacred and profane, since the formal paradox, conventionally regarded as low, parodies at the same time as it imitates the divine act of Creation. And yet, who can accuse the paradoxist of blasphemy, really? Since his subject *is* nothing, he cannot be said to be impious in taking the Creator's prerogative as his own – for nothing, as all men know, can come of nothing. Nor indeed is he directing men to dangerous speculation, since at the very most he beguiles them into – nothing. And most important of all . . . if the paradoxist lies, he does not lie, since he lies about nothing."[45]

The two most common trends of this sort are to be found in the 'all or nothing' paradoxes and the amusing penchant for double entendres about

Nothing displayed by writers and playwrights. Poets joined in the game as well, with works like *The Prayse of Nothing*:[46]

> "Nothing was first, and shall be last
> for nothing holds for ever,
> And nothing ever yet scap't death
> so can't the longest liver:
> Nothing's so Immortall, nothing can,
> From crosses ever keepe a man,
> Nothing can live, when the world is gone,
> for all shall come to nothing."

and *On the Letter O*,

> "But O enough, I have done my reader wrong
> Mine O was round, and I have made it long."[47]

or Jean Passerat's *Nihil*, informing us that

> "Nothing is richer than precious stones and than gold; nothing is finer than adamant, nothing nobler than the blood of kings; nothing is sacred in wars; nothing is greater than Socrates' wisdom – indeed, by his own affirmation, nothing *is* Socrates' wisdom. Nothing is the subject of the speculations of the great Zeno; nothing is higher than heaven; nothing is beyond the walls of the world; nothing is lower than hell, or more glorious than virtue."[48]

and so on, and on, and on.

These word games soon become a little tedious to our ears. They had the goal of generating lots of words from nothing by means of talking about Nothing. For a time, the genre was a fashionable form of philosophical nonsense verse. Several paradoxical juxtapositions occur again and again. There is the picture of the circle representing, on one hand zero, and, on the other, the encompass of everything. There is the egg, shaped like zero but promising to become the generator of new life. It was pregnant with creativity just like the mathematicians' zero, waiting to be added to other figures to create larger numbers. And in the background there is the sexual allusion to the circle which represents the female genitalia. This is a running joke in Elizabethan comedies although much of the humour is lost on us. Fortunately, there is a famous example of this genre that is widely known and appreciated. It is intriguing because it shows that the paradoxes and puns of Nothing attracted the interest of the greatest of all wielders of words.

SHAKESPEAREAN NOTHINGS

"Is this nothing?
Why, then the world and all that's in't is nothing;
My wife is nothing: nor nothing have these nothings,
If this be nothing."

William Shakespeare, The Winter's Tale[49]

Shakespeare was much taken with all the linguistic and logical paradoxes of Nothing. For good measure he entwined them with the double entendres of the day to add yet another dimension to the many-layered works that are his hallmark. The comedy *Much Ado About Nothing* is a wonderful example[50] of the deftness with which games could be played with words that others had struggled to enliven. First appearing in print in 1600, and probably written during the preceding two years, the title of this play immediately illustrates the general fascination with the ambiguities of Nothing that were in vogue in Shakespeare's time.[51] In the fourth act, the prospective lovers Beatrice and Benedick use the ambiguities of Nothing as a subtle smokescreen so that each hearer can choose to interpret Nothing in a positive or a negative way:

"Benedick: I do love nothing in the world so well as you.
Is not that strange?

Beatrice: As strange as the thing I know not. It were as possible for me to say I loved nothing so well as you. But believe me not; and yet I lie not. I confess nothing, nor I deny nothing."[52]

Shakespeare plays upon other dimensions of Nothing as well. In the tragedies *Hamlet* and *Macbeth* we find the philosophical and psychological paradoxes of Nothing deeply interwoven with human experience. Macbeth is repeatedly confronted with the paradoxes of Nothing and the horrors of non-Being: he despairs that

"Nothing is
But what is not."[53]

Hamlet explores how Nothing can have paradoxical meaning and content. In contrast to Macbeth, who rails that

"Life's . . . a tale
Told by an idiot, full of sound and fury,
Signifying nothing",[54]

the Prince of Denmark finds consolation in death and convoluted speculation about Nothing. They stand in stark contrast about what it means to . be and not to be, for

"where Macbeth discovers that death is oblivion, Hamlet discovers that it is not. Macbeth discovers that, when death is oblivion, life is insignificant. Hamlet discovers that when one does not fear death, life with all its painful responsibilities can be borne and even borne nobly. In the end Hamlet knows for himself the relation between 'to be' and 'not to be' by which even his own death can affirm life."[55]

Yet even Hamlet makes full use of the double entendres associated with Nothing and the female form in this exchange with Ophelia:[56]

> "Hamlet: Lady, shall I lie in your lap?
> Ophelia: No, my lord.
> Hamlet: I mean, my head upon your lap?
> Ophelia: Ay, my lord.
> Hamlet: Do you think I meant country matters?
> Ophelia: I think nothing, my lord.
> Hamlet: That's a fair thought to lie between maids' legs.
> Ophelia: What is, my lord?
> Hamlet: Nothing."

In *King Lear*, Shakespeare tells of the destruction of Lear by all that emanates from Nothing. The play has a recurrent theme of quantification, numbering and reduction. Two of Lear's daughters make pretentious statements of love and respect for him in return for parts of his kingdom, but the third, Cordelia, will not play this cynical game or just remain silent. Her encounter with her father introduces a typical play on Nothing:[57]

> "Lear: . . . what can you say to draw
>   A third more opulent than your sisters'? Speak!
> Cordelia: Nothing, my lord.
> Lear: Nothing?
> Cordelia: Nothing.
> Lear: Nothing will come of nothing. Speak again."

From this ominous beginning many are reduced to Nothing. Cordelia is hanged. Lear's Fool asks him 'Can you make use of nothing?' and Lear repeats his admonition to Cordelia, 'Why, no, boy. Nothing can be made out of nothing.' But the Fool responds by reducing Lear to Nothing:

> "Thou art an O without a figure. I am better than thou art now; I am a fool, thou art nothing."

Lear's other daughters, Goneril and Regan, reduce Lear to zero in more practical ways, demanding that he reduce the size of his entourage, halving and halving it until there is only one left and then Regan asks, 'What need one?' Lear shows Shakespeare[58] grappling with the double meanings of Nothing, the metaphysical void and the end result of taking away what one has, bit by bit, if one exports the arithmetic of buying and selling into the human realms of love, loyalty and duty. Things then don't always add up. Madness is not far away. On that you can count.

Shakespeare explored all the meanings of Nothing: from the simplicity of zero, the nonentity of the cipher, the emptiness of the void, and the absence of everything it witnessed, to the contrast between the whole and the hole that was zero, the circle and the egg, hell, oblivion and the necromancer's circle. His explorations can be roughly divided into those that pursue the negative aspect of Nothing and those that pursue the positive. On the negative side we see the focus on the absence of things, on denial, apathy and silence. These invariably bring bad consequences and reveal some of the awful results of meaninglessness. By contrast, the positive side of Nothing lays stress on the power of Nothing to generate something. Just as zero lay at the beginning of an ever-increasing sequence of numbers, so the sexual connotations of Nothing and the pregnant power of the egg symbolised fruitfulness and multiplication, the growing of something out of nothing. Indeed, it was just this multidimensional proliferation that Shakespeare's own work displayed.[59]

One should not think that the linguistic gymnastics of nihil paradoxes are a thing of the past. While it is not common for these word games to be played by modern writers, they can still be found if you know where to look. Here is Jean-Paul Sartre trying to convey information about the origin of negation:

> "Nothingness is not, Nothingness is 'made-to-be', Nothingness does not nihilate itself; Nothingness 'is nihilated' . . . It

would be inconceivable that a Being which is full positivity should maintain and create outside itself a Nothingness or transcendent being, for there would be nothing in Being by which Being could surpass itself towards Non-Being. The Being by which Nothingness arrives in the world must nihilate Nothingness in its Being, and even so it still runs the risk of establishing Nothingness as a transcendent in the very heart of immanence unless it nihilates Nothingness in connection with its own being. The Being by which Nothingness arrives in the world is a being such that in its Being, the Nothingness of its Being is in question. The being by which Nothingness comes to the world must be its own Nothingness . . ."[60]

and so on, for more than 600 pages.

PARADOX LOST

"What did the mystic say to the hot-dog vendor?
Make me one with everything."

Laurence Kushner

By the end of the seventeenth century the literary fascination with the paradoxes of Nothing had run its course.[61] It ceased to be a mainspring of imaginative exploration in both literature and philosophy. Writers simply mined out the seam of possibilities and moved on to explore new ideas. Philosophers came to distrust these games with words and they were seen increasingly as mere puzzles to amuse. They were no longer considered to provide a route into deep truths about the nature of things. The increasing stress placed upon observation and experiment relegated the paradoxes of Nothing to a linguistic backwater from which they would not reappear until the beginning of the twentieth century. The sea change in attitudes is displayed in Galileo's *Dialogue Concerning Two World Systems*,[62] which includes a discussion of the dangers of treating the contemplation of 'words' as a superior route to truth than the study of 'things'. Simplicio cautions that 'everybody knows that you may prove whatever you will' by means of linguistic paradoxes. Galileo equated 'paradox' with vague, unverifiable word games that had no place in the development of science, which was typified by the logic of testable chains of cause and effect. For example, the famous 'Liar paradox', credited to Epimenides, which St Paul repeats, that 'all Cretans are liars, one of their own poets has said so' was condemned as 'nothing but a sophism ... a forked argument ... And thus, in such sophisms, a man may go round and round for ever and never come to any conclusion.'

Galileo had the highest regard for mathematical knowledge of the world. He recognised that our knowledge of most things was necessarily imperfect. We can only know as much as Nature reveals to us, but in the field of mathematics we have access to a part of the absolute truth at the heart of things. For

> "the human intellect does understand some propositions perfectly, and thus in these it has as much absolute certainty as Nature itself has. Of such are the mathematical sciences alone; that is, geometry and arithmetic, in which the Divine intellect indeed knows infinitely more propositions, since it knows them all. But with regard to those few which the human intellect does understand, I believe that its knowledge equals the Divine in objective certainty."[63]

This remarkable passage shows how mathematics and geometry came to support the belief that it is possible for humans to know some of the absolute truth of things. Because Euclid's geometry was believed to be true – a precise description of reality – it provided important evidence that human thought could penetrate the nature of ultimate truth in at least one area. And, if it could do this in the realm of mathematics, then why not in theology too? Paradoxes were not part of this domain of ultimate reality. Ironically, in the twentieth century Kurt Gödel would turn these beliefs on their head in a striking way. Gödel showed that there are statements of arithmetic that can be made using the rules and symbols of arithmetic which it is impossible to show to be either true or false using those rules. The golden road to truth that Galileo loved must always give rise to statements that are unverifiable. Remarkably, Gödel established this extraordinary truth about the limits of mathematics by taking one of the linguistic paradoxes that Galileo rejected and transforming it into a statement about mathematics. But long before Gödel's work, the absolute truth of mathematics had been undermined. Mathematicians of the nineteenth century had shown that Euclid's classical geometry was but one amongst many. There were an infinite number of possible geometries, each obeying their own set of self-consistent axioms, different from Euclid's. These new geometries described lines and figures drawn on curved surfaces rather than the flat ones that Euclid assumed. None of these systems was any 'truer' than any of the others. They were each logically consistent, but different, axiomatic systems. None of them had any special claim to be part of the absolute truth at the heart of things. Later, this 'relativism' would spread even to logic itself. The simple logic of Aristotle was revealed to be but one system of reasoning amongst an unlimited catalogue of possibilities.

The Galilean distinction between the quagmire of paradox and the sure path of science paved with conjectures and refutations was an important one. It moved science towards the modern era of experimental investigation. No longer were important questions solved by recourse to authorities like Aristotle.[64] Human self-confidence was reawakened. It was possible to do better than the ancients. And one did not have to be more inspired to do so. A superior method was what was needed: look and see. If the question was whether or not there could be moons around the planet Jupiter the answer was not to be found by philosophical arguments about the appropriateness of this state of affairs or the natural places for moons to reside, it could be decided by just looking through a telescope.

In this chapter we have traced the fate of Nothing in the hands of philosophers and writers with very different aims. Medieval scholars inherited the world pictures of the Greeks and the mathematical systems of the Far East. Both had distinctive pictures of Nothingness etched into their fabrics. The need to handle the philosophical and theological implications of Nothing was in many ways fuelled by the acceptance of the idea in simple mathematics, where it proved uncontentious and useful. It replaced nothing and it could exist merely as a sign that signalled the divide between profit and loss, prosperity and ruin. It was a symbol with a prosaically positive message. The books balanced; nothing was missed out; all debts were repaid. These were the messages that the zero symbol sent throughout the world of business. Away from the world of numbers there were bigger issues at stake. Nothing was entwined with theological issues of the greatest consequence. Was it the realm from whence the world was made? And if it was, how could it not be something? We are content with the cogency of nothing at all, so long as we do not pursue the idea too closely. But there was an influential Greek view of the nature of things which made the whole concept of nothing at all quite incomprehensible. Plato's explanation for things saw them as manifestations of the eternal forms behind the appearances. Even if there were no things, no expressions of those eternal forms, the blueprints themselves must always exist. If they didn't then there would be no way in which the appearance of the world could be in-formed. The eternal forms were the source of the in-formation required to turn the potential into the actual. Nothing was no part of either.

One of things that we have seen in the struggle to make sense of the vacuum and its possible reality is a medieval willingness to conduct experiments, both thought-experiments which appeal to common experience and more contrived sequences of events which demand careful observation and interpretation. This appeal to the behaviour of the world as a source of reliable knowledge did not begin with Galileo, but with him it started to become the only trusted guide to the truth behind everyday things. This was not so much because other guides were mistrusted, merely that they were so hard to interpret clearly and reliably. The medieval philosophers like Bacon and Burley began a tradition of inquiry and a search for the vacuum that would be taken up by Galileo and his contemporaries with a brilliant acuteness. Nothing better displays the phase transition from natural philosophy to natural science.

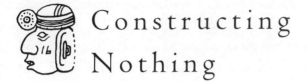

chapter three

Constructing
Nothing

"On the empty desk sat an empty glass of milk."

BBC Radio 3[1]

THE SEARCH FOR A VACUUM

"Nature, it seems, is the popular name
for milliards and milliards and milliards
of particles playing their infinite game
of billiards and billiards and billiards."

Piet Hein, *Atomyriades*

While writers like William Shakespeare were plumbing the depths of the moral vacuum, others were seeking to create nothing less than a real physical vacuum. For more than two thousand years philosophers had argued fervently about the reality of a physical vacuum: the possibility that there could be a region of space that contains absolutely nothing. Both Aristotle and Plato denied, for quite different reasons, that such a vacuum could exist but other ancient thinkers disagreed. The Roman philosopher Lucretius was convinced that matter was composed of small constituent particles, which we would call 'atoms', and that the basic nature of the Universe was a motion of these atoms in the void that lay between them.

This picture of Nature, that we now call atomism,[2] led its seventeenth-century supporters to countenance the existence of a vacuum in situations that were amenable to experimental investigation. Nor was it quite so mysterious as the theologians had claimed. It could be envisaged as the endpoint of a sequence of mechanical processes that sucked the contents out of a jar. As more of the contents were extracted so the closer did the inside of the jar come to resembling one which could be said to contain nothing at all. Of course, from the perspective of a sceptical philosopher this experiment might appear a little oversimplified. Even though all the air might be removed from the jar its interior could not be said to contain nothing. It was still subject to the laws of Nature. It remained part of the universe of space and time. One could still argue with justification that a perfect vacuum could never be created. For the pragmatist this claim would be supported by the manifest impossibility of extracting every last atom from the jar. For the natural philosopher a last-ditch defence was still available by appeal to a subtle distinction between jars that were completely empty and jars that were merely empty of everything of which they might be emptied. Nevertheless, the ensuing search for a physical vacuum was visually dramatic and it changed for ever the question of the character of that vacuum. It was now to become primarily a scientific question to which there were scientific answers.

The most fruitful investigations of the vacuum were conceived by seventeenth-century scientists investigating the behaviour of gases under pressure. If a container was to be evacuated of its contents, then the only way to get all the air out of the container was by sucking it out. This required the creation of a pressure difference between the inside and the outside of the container. A pump was needed and such devices existed for the pumping of water on ships and farms. In 1638 Galileo wrote[3] that he had noticed that there was a limit to how high he could pump water using a suction pump. It would rise by ten and a half metres but no higher. He tells us about the problem of trying to pump water up from a cistern when its level had fallen too low:

"When I first noticed this phenomenon I thought the machine was out of order; but the workman whom I called in to repair it told me the defect was not in the pump but in the water which had fallen too low to be raised through such a height; and he added that it was not possible, either by a pump or by any other machine working on the principle of attraction, to lift water a hair's breadth above eighteen cubits; whether the pump be large or small this is the extreme limit of the lift."

Evidently Galileo was far from being the first to notice this irritating fact of agricultural life. There must have been farm workers and labourers trying to siphon water out of flooded trenches all over Europe who had come to appreciate it the hard way. Consequently there was good reason to devise suction pumps which could overcome this limit. As these machines improved they stimulated scientists to investigate why they worked at all. They were led to appreciate that if air could be removed from a closed space then the evacuated region would tend to suck things into it. At first this appeared to confirm Aristotle's ancient precept that 'Nature abhors a vacuum': create an empty space and matter will move so as to refill it. Yet Aristotle maintained that this happened because of a teleological aspect to the working of the world. He expected matter to be drawn to fill the vacuum because it had that end in view. This is quite different from the type of explanation sought by Galileo. He was seeking a definite cause or law of Nature that would predict the future from the present physical state of affairs.[4] Galileo saw that there was something unsatisfactory about using the inability of water pumps to raise water above some definite height as evidence for Nature's abhorrence of a vacuum. For why did Nature's level of abhorrence reach such a height ('eighteen cubits') and no further?

Galileo's interest in the vacuum was not really philosophical. He was content to believe that it was impossible to make a true vacuum. For his purposes it was enough to produce a region that was almost empty. The reason for his interest in such a region is not hard to find. His deep insights into the behaviour of bodies falling under gravity had led him to recognise that air resistance played an important role in determining how things would fall under the pull of gravity. If objects of different mass, or of different size, are dropped simultaneously in a vacuum (where there is no air resistance to impede their fall to the ground), then they should experience the same acceleration and reach the ground at the same moment. Legend has it that Galileo performed this experiment by dropping objects from the Leaning Tower of Pisa, but historians regard this as rather unlikely to have been the case. However, in reality, in the Earth's atmosphere a stone and a feather certainly do not hit the ground simultaneously if released together because of the very different effects of air resistance upon them. By producing a good vacuum, Galileo could get a better approximation to the true vacuum where his idealised laws of motion were predicted to hold exactly. In fact, this experiment with a falling feather and a rock was one of the first things that was done by the first Apollo astronauts to walk on the Moon for all to see on television. In the absence of an atmosphere to resist their motion, the two objects hit the ground together, just as Galileo predicted. This type of experiment was first carried out in less ideal conditions by the French scientist, Desaguliers, in 1717, as a demonstration for Isaac Newton at the Royal Society in London. Instead of a feather and a stone he used a guinea coin[5] and a piece of paper. The *Philosophical Transactions of the Royal Society* reported that

> "Mr Desaguliers shew'd the experiment of letting fall a bitt of Paper and a Guinea from the height of about 7 foot in a vacuum he had contrived with four glasses set over one another, the junctures being lined with Leather liquored with Oyle so as to exclude the Air with great exactness. It was found that the paper fell very nearly with the same Velocity as the Guinea so that it was concluded that if so great a Capacity could have been perfectly exhausted, and the Vacuum preserv'd, there would have been no difference in their time of fall."

The puzzle of the water pumps was solved in 1643 by one of Galileo's students, Evangelista Torricelli, who worked as his secretary in 1641–2 and eventually succeeded him as the court mathematician to the Tuscan Grand Duke Fernando II, a post he held until his premature death in 1647, when aged only thirty-nine. Torricelli realised that the Earth's atmosphere carried a weight of air which bore down on the Earth and exerted a pressure at its surface. It was this 'atmospheric pressure' that he suspected, but could not rigorously prove, was the real reason why air tended to fill up any vacuum that we try to create. Using water was a cumbersome (although cheap) way to carry the investigations further. Eighteen cubits is about 10.5 metres and this is a tall order to study in a laboratory. But if he could use a liquid that was much denser than water, then the maximum height it could be pumped would be smaller. The densest liquid of all is the liquid metal, mercury. It is almost fourteen times denser than water and so we would expect that the maximum height it could be raised would be fourteen times less than that for water, giving a convenient height of just 76 centimetres of mercury. Using mercury, Torricelli[6] constructed the first simple manometer without even needing a pump to raise the mercury, as shown in Figure 3.1.

He took a straight glass tube that was longer than 75 centimetres, sealed at one end by the glassblower but left open at the other. Using a bowl of mercury he filled the tube right to the top, sealed the open end with his finger, and then inverted the tube to stand upright with its open end under the surface of the mercury in the bowl (see Figure 3.1). When he removed his finger, the mercury level dropped down the tube. Every time you do this experiment at sea level, no matter how wide the tube, the level taken by the mercury is approximately 76 centimetres above the surface of the mercury in the bowl.[7]

The remarkable thing about Torricelli's experiment was that for the first time it appeared to create a sustained physical vacuum. When the tube was first filled with mercury there was no air within it. Yet after the tube had been inverted the mercury fell, leaving a space in the sealed tube above it. What did it contain? No air could get in. Surely it must be a vacuum. On

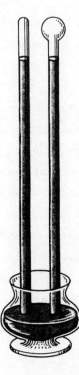

Figure 3.1 *Two examples of Torricelli's barometer.*[8] *The column of mercury in each vertical tube is balanced by the pressure of the atmosphere on the surface of the mercury in the dish. At sea level the height is about 76 cm.*

11 June 1644 Torricelli wrote to one of his friends, Michelangelo Ricci, revealing some of his thoughts about the profound implications of his simple experiment:[9]

> "Many people have said that it is impossible to create a vacuum; others think it must be possible, but only with difficulty, and after overcoming some natural resistance. I don't know whether anyone maintains that it can be done easily, without having to overcome any natural resistance. My argument has been the following: If there is somebody who finds an obvious reason for the resistance against the production of a vacuum, then it doesn't make sense to make the vacuum the cause for these effects. They obviously must

depend on external circumstances . . . We exist on the bottom of an ocean composed of the element air; beyond doubt that air does possess weight. In fact, on the surface of the Earth, air weighs about four hundred times less than water . . . the argument that the weight of air such as determined by Galileo is correct for the altitudes commonly inhabited by man and animals, but not high above the mountain peaks; up there, air is extremely pure and much lighter than the four hundredth part of the weight of water."

The reason for the behaviour of the column of mercury in Torricelli's tube is that the force exerted by the weight of air in the atmosphere above the bowl of mercury acts on the surface of the mercury and causes the mercury to rise up the tube to a level at which its pressure balances that exerted by the air on the surface of the mercury bowl. Actually, the height of the mercury column is only approximately equal to 76 centimetres. It varies as the weather conditions change and from place to place on the Earth's surface. These changes reflect the change in atmospheric pressure created by the winds and other changes in the density of the atmosphere that are produced by variations in temperature. When we see a weather map in a newspaper or on the television it will display isobars which trace the contours of equal pressure. These effects of weather on the pressure exerted by the atmosphere allowed Torricelli's device to provide us with the first barometer. We notice also in his account to Ricci that he has realised that the result of his experiment depends upon the altitude at which it is conducted. The higher one climbs, the less atmosphere there is above and the lower the air pressure weighing down on the mercury column.

Torricelli was a talented scientist with many other interests besides air pressure. He determined laws governing the flow of liquids through small openings and, following in the footsteps of his famous mentor, deduced many of the properties of projectile motion. Not merely a theorist, he was a skilled instrument maker and lens grinder, making telescopes and simple microscopes with which to perform his experiments, and he made a considerable amount of money by selling them to others as well.

Torricelli's simple experiment led eventually to the acceptance of the radical idea that the Earth was cocooned in an atmosphere that thinned out as one ascended from the Earth's surface and was eventually reduced to an empty expanse that we have come to call simply 'space' or, if we keep going a bit further, 'outer space'. This dramatic background stage for life on Earth provided the beginning for many reassessments of humanity's place and significance in the Universe. Copernicus had published his startling claims that the Earth does not lie at the centre of the solar system about one hundred years before Torricelli's work. The two are closely allied in spirit. Copernicus moves us from a central location in the Universe while Torricelli reveals that we and our local environment are made of a different density of material than the Universe beyond. We are isolated, swimming in a vast emptiness. Later, we shall find that this emptiness of space has remarkable consequences for us and for the possibility of life in the Universe.

Spurred on by Torricelli's demonstrations and suggestions, other scientists around Europe started to investigate the empty space at the top of the mercury column, to discover its hidden properties, subjecting it to magnets, electric charge, heat and light. Robert Boyle[10] in England used simple 'vacuum pumps' constructed by Robert Hooke to evacuate much larger volumes than those naturally produced by Torricelli and studied what happened to mice and birds placed in jars as they were gradually evacuated of air.[11] He appears to have escaped the attentions of the seventeenth-century equivalent of the Animal Liberation Front.

Boyle was extremely wealthy. His family were substantial Irish landowners in County Waterford. His serious study of science began in earnest after graduating from Eton in 1639 when he first read Galileo's works whilst making a grand European tour with his private tutor. Upon his return he established himself in Dorset and began his impressive experimental scientific work. Later, he would move to Oxford and become one of the founding fellows of the Royal Society. Boyle had no need to seek grant support. He inherited a large fortune which allowed him to buy expensive pieces of scientific equipment and hire skilled technicians to help maintain and modify them. Boyle sought to exorcise the notion that the vacuum at the top of Torricelli's barometer possessed a suction that was drawing the mercury up the tube in accord with traditional Aristotelian beliefs about the tendency for Nature to remove a vacuum. Such a notion was not held without reason. If you put your finger over the end of a glass tube, it did feel as if it was being slightly sucked up into the tube because it was difficult to remove it. Boyle laid the foundations for a straightforward explanation for the height of the mercury in terms of the difference in pressure between the atmosphere and the 'vacuum' inside the tube. Rival Aristotelian theories proposed that there was an invisible ropelike structure, called a *funiculus* (from the Latin *funis*, for rope), which pulled on the mercury, preventing it falling to the bottom of the tube. Boyle was able to demonstrate the superiority of the air pressure theory by using it successfully to predict the level attained by the mercury when the outside pressure was changed to different values.[12]

The most spectacular experiment inspired by Torricelli's work was conducted in 1654 by Otto von Guericke,[13] a German scientist who for thirty years was one of the four mayors of the German city of Magdeburg (Figure 3.2).

This civic status was of great help to him in making a memorable public display of the reality of the vacuum. His celebrated 'Magdeburg Hemispheres' demonstration involved carefully building two hollow bronze hemispheres which fitted closely together to form a good seal. A pump was requisitioned from the local fire service and attached to a valve on one of them so that the air could be sucked out after they were joined together to form a spherical shell. After much pumping Von Guericke announced to his audience that he had created a vacuum. Moreover, Nature was rather happy with it. Far from shunning or trying to remove it, as the ancients were so fond of preaching, Nature strenuously defended the vacuum against any attempt to destroy it! Just so that no one could miss the point, two teams of eight horses were harnessed together and hitched up to each hemisphere and then driven off in opposite directions in order to tear the hemispheres apart. They failed! Then Von Guericke opened the valve to let the air back in and the hemispheres could be effortlessly separated. They don't do experiments like that any more! Actually, the two teams of

Figure 3.2 *Otto von Guericke.*[14]

eight horses proved rather hard to handle and required six trials before he could get each team member pulling in the same direction at the same time. The two Magdeburg hemispheres can still be seen in the Deutsches Museum in Munich (Figure 3.3).

The result of all this work was to convince scientists that the Earth was surrounded by a substantial body of air which exerted a significant pressure on its surface. By carefully studying its effects, it was possible to explain all

Figure 3.3 *The Magdeburg Hemispheres Experiment.*[15]

sorts of behaviours of gases and liquids in detailed mechanical terms rather than merely ascribing them to the vague notion that 'Nature abhors a vacuum' as the ancients did. Historians of science have highlighted the mundane study of air pressure as a turning point in our study of Nature; teleological notions of the 'inclinations' in the natural order of things brought about by mysterious occult forces were superseded by explanations that used only the concepts of matter and motion.

Von Guericke was a practical engineer of great ingenuity. He both liked and invented machines. But he was still fascinated by the ancient philosophical questions about the reality of the vacuum and their implications for the Christian doctrine of the creation of the world out of nothing. In his account of his experimental investigations he devotes a substantial section[16] to airing his views about the void, which have a strong affinity to the medieval scholastic ideas that formed the philosophical tradition in which he worked. Von Guericke's book is rather overblown. He has something to say about just about everything under the sun, and a good deal about things above it as well. His position in local government ensured that there would be an effusive dedication by one of the local noblemen. Indeed, Johannes von Gersdorf is moved to poetry in his tribute to the experimenter of Magdeburg, 'the most distinguished and excellent gentleman, Otto von Guericke':

> "To delve into the manifold mysteries of nature
> is the task of an inquiring and fertile mind.
> To follow the tortuous paths of nature's wondrous ways
> is work more difficult and not designed for everyone.
> You, Distinguished Sir, Magdeburg knows as its
>   Burghermaster
> as well as an outstanding researcher in the field of science.
> Whether one speaks with you informally or studies your
>   work alone,
> he will soon confirm your genius openly and without a
>   feeling of doubt.
> May I make a small joke? While you prove quite clearly
>   that a vacuum exists
> in your Book, there is not a vacuum to be seen!"

For Von Guericke everything that existed could be put into one of two classes: it was either a 'created something' or an 'uncreated something'. There could be no third way: no class that we can call 'nothing'. Since 'nothing' is the affirmation of something and the opposite of something else, it must be a something. Thus it falls into the category of either the 'created somethings' or the 'uncreated somethings'; or maybe, he feels, 'nothing' has a call on belonging to both categories. Thus an imaginary animal like a unicorn is nothing in the sense of being non-existent; that is, it is not a thing. But because it exists as a mental conception it is not absolutely nothing. It has the same type of existence as a human thought. Thus it qualifies as a created something. Von Guericke wanted to view the Nothing that was before the World was made as an uncreated something, so that he could say that before the World was created there was Nothing, or equally, there was an uncreated something. In this way he guards against sounding like a heretic.

Von Guericke summarised his lyrical philosophy of the void in a great psalm in honour of Nothing (*Nihil*). It gives a flavour of thinking that one would not have immediately associated with down-to-earth experimental demonstrations that the vacuum could be controlled by air pumps. It is worth reading at length. He joins various concepts of Nothing, empty space and imagined space together into one and the same concept, for

> "everything is in Nothing and if God should reduce the fabric of the world, which he created, into Nothing, nothing would remain of its place other than Nothing (just as it was before the creation of the world), that is, the Uncreated. For the Uncreated is that whose beginning does not pre-exist; and Nothing, we say, is that whose beginning does not pre-exist. Nothing contains all things. It is more precious than gold, without beginning and end, more joyous than the perception of bountiful light, more noble than the blood of kings, comparable to the heavens, higher than the stars, more powerful than a stroke of lightning, perfect and blessed in every way. Nothing always inspires. Where nothing is, there ceases the jurisdiction of all kings. Nothing is without any mischief. According to Job the Earth is suspended over Nothing. Nothing is outside the world. Nothing is everywhere. They say the vacuum is Nothing; and they say that imaginary space – and space itself – is Nothing."[17]

Von Guericke believed that space was infinite and likely to be populated by many other worlds like our own. He used the idea of the infinite world to support his argument that there is really no difference between real and imagined space. For, although we might think that unicorns inhabit only an imaginary space, he argues that if space is infinite then we are reduced to imagining some of its properties, just like we are the unicorns. In fact, Von Guericke equated infinite space, or Nothing, the uncreated something, with God.

A TALE OF TWO NOTHINGS

"It is hard to think of any modern parallel to the shiver of horror engendered by the mere suggestion to a man of the seventeenth century that a vacuum could effortlessly exist and be maintained; a materialist forced to admit irrefutable evidence of life after death might offer a fair analogy."

Alban Krailsheimer[18]

It is well to remember that these great experiments with air pressure focused attention on the problem of two Nothings. There was the abstract, moral or psychological 'nothing', juggled with by playwrights and philosophers. It was a Nothing entirely metaphysical in nature; one that you didn't have to worry about if you didn't want to. It could stay as poetry. Set in stark prosaic contrast was the problem created by the attempts to create a real physical vacuum in front of your very eyes by evacuating glass tubes or metal hemispheres. This vacuum exerted forces and could be used to store energy. This was a very useful Nothing.

The seventeenth-century thinker who did most to join the two conceptions together was a polymath with diverse, seemingly contradictory, interests who liked to engage with problems that possessed a hint of the impossible or the fantastic. Some of those interests were physical, some were mathematical, while others were entirely theological. Blaise Pascal was born in the French town of Clermont in 1623. He died only thirty-nine years later, but in that short space of time he laid the foundations for the serious study of probability, constructed the second mechanical calculating machine, made significant discoveries about the behaviour of gases under pressure, and found new, important results in geometry and algebra.

Finally, his most famous work was his unfinished collection of fragmentary 'thoughts', the *Pensées*,[19] that remained incomplete at the time of his early death. All this was achieved from unpromising beginnings. Pascal's mother died when he was just three years old, leaving the young boy at the mercy of his father's theories of education. They moved to Paris where Etienne, a successful lawyer, decided to educate his rather sickly son himself in isolation from other children. Although Pascal Senior was an able mathematician, he was determined that his son should not study mathematics until he reached the age of fifteen, and all mathematics books were removed from their house. Not content with the diet of Latin and Greek that resulted, the young Pascal gradually came in contact with friends of his father who shared an interest in mathematics. Not to be denied, he evaded his father's educational restrictions by rediscovering a number of geometrical properties of triangles for himself at the age of twelve. Surprised, his father relented and gave him a copy of Euclid's book of geometry to work with. Soon afterwards the family uprooted and moved to Rouen where his father had been appointed tax collector for the region. The young Pascal prospered there. At the age of sixteen he presented his first mathematical discoveries of new theorems and geometrical constructions to a regular meeting of Paris mathematicians convened by Mersenne, one of the most notable number theorists of the day. His first published work, on geometry, appeared just eight months later. So began Pascal's career of invention and discovery (see Figure 3.4).

Pascal's interest in air pressure and the quest to create a perfect vacuum began in 1646. Unfortunately, this work threw him on to a collision course with the views of René Descartes, the most influential French natural philosopher of his day. In Rouen, Pascal came to hear of Torricelli's remarkable experiments, conducted a few years before. Teaming up with Pierre Petit, a fortifications engineer and friend of his father, he began a series of telling experiments.[20] The most important was planned by Pascal and carried out by his brother-in-law, Florin Périer. It sought to demonstrate the claims that Torricelli was making in his letter to Ricci about the thinning out of the Earth's atmosphere at high altitude. On 15 November

Figure 3.4 *Blaise Pascal.*[21]

1647 Pascal wrote to Périer asking him to compare the mercury levels in a Torricelli tube at the base and at the summit of a local mountain:

> "if it happens that the height of the quicksilver [mercury] is less at the top than at the base of the mountain (as I have many reasons to believe it is, although all who have studied the matter are of the opposite opinion), it follows of necessity that the weight and pressure of the air is the sole cause of this suspension of the quicksilver, and not the abhorrence of the vacuum: for it is quite certain that there is

much more air that presses on the foot of the mountain than at its summit."[22]

After a delay of several weeks due to bad weather, Périer gathered together his team in the convent garden in the town of Clermont. He prepared two identical tubes of mercury, adopting the same method as Torricelli had used. The heights of the mercury columns were measured carefully in the presence of an audience of local worthies and verified to be identical. The local priest was then left in charge of one of the mercury columns while Périer's team headed for the summit of Mount Puy-de-Dôme in the Auvergne, 1465 metres above sea level. They read the height of mercury at different altitudes on their ascent and at the summit itself. Mission accomplished, they returned to the convent to check that the height of mercury in their instrument was the same as the one left in the priest's care. It was. The change in level between the convent and the mountain top was a clear 8.25 centimetres.

What these measurements established for the first time, on 9 September 1648, was that the pressure of air decreased as one ascended the mountain. This result had a tremendous impact on all involved. Pascal wrote that the experimenters 'were carried away with wonder and delight'. The fact that the effect which they discovered was so large inspired one of them, Father de la Mare, to look for the difference in mercury level when he took a barometer from ground level to the top of the 39-metre-high tower of the cathedral of Notre Dame de Clermont. The difference was 4.5 millimetres: small but still quite measurable. When Pascal heard the result he repeated the experiment in the tallest buildings in Paris, finding a similar measurable trend: the taller the building the bigger the pressure drop at the top. He soon realised that sensitive measurements of the variation of pressure with altitude could be used to determine altitude if a detailed enough understanding of the correlation between atmospheric pressure and altitude could be obtained independently. In the 350 years since Pascal's first measurements we have built up a detailed picture of the Earth's tenuous atmosphere (see Figure 3.5). Subsequently, he discovered that the baromet-

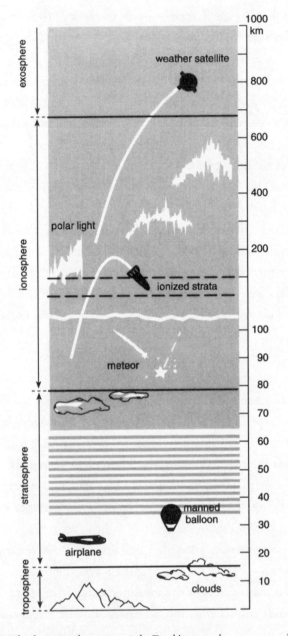

Figure 3.5 *The change in the nature of the Earth's atmosphere up to an altitude of 1000 kilometres above sea level.*[23]

ric level at one place could change with the weather conditions, a fact that our modern use of the barometer exploits. A modern weather map, laced with isobars, is shown in Figure 3.6.

Pascal maintained that the empty space at the top of the mercury tube was a real vacuum. Pascal's opponents were not especially interested in the practical implications of pneumatics and hydraulics, but they were concerned about the philosophical implications of such a claim, not least because they were being made by a young man of twenty-three who possessed no formal academic qualifications, merely an extremely stubborn and persistent character. In Italy, Torricelli's group worked on many experiments between 1639 and 1644 but they did not pursue their researches any further, probably for fear of opposition by the Church – Giordano Bruno was burned at the stake in 1600 and Torricelli's mentor, Galileo, remained under house arrest by the Inquisition in the nine years before his death in 1642. But Pascal, encouraged by Mersenne, who had learned of the experiments carried out in Rome, held no such fears despite his deeply religious inclinations. He improved and

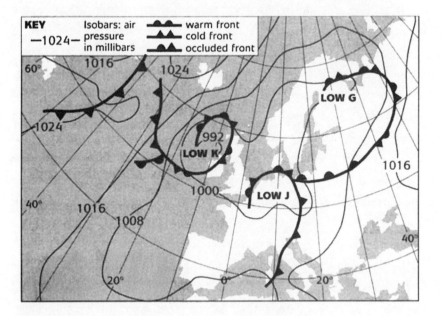

Figure 3.6 *A weather map showing isobars, contour lines of equal atmospheric pressure. Winds blow from high to low pressure areas.*[24]

extended Torricelli's experiments in many ways. He experimented with water and red wine and various oils, as well as mercury. These required large, often spectacular, experiments with long tubes and huge barrels to be performed in the streets of his home town. This showmanship did not endear him to his conservative opponents.

Pascal faced opposition to the recognition of the reality of a physical vacuum on two fronts. The traditionalist Aristotelians had long exercised a strong influence on physics and they denied the possibility of making a vacuum. They explained the observed changes in Nature by 'tendencies' which really explained nothing at all; things grew because of a life force, objects fell to earth because of a property of heaviness. This was just a name game that could make no decisive predictions about what would happen in situations never before observed. But the Aristotelians were not the only ones to deny the possibility of a vacuum. The Cartesians, followers of Descartes, followed a unified natural philosophy which attempted to deduce the behaviour of the physical world in mathematical terms, by means of specific universal laws. However, this modern-sounding initiative did not allow space to be discussed without matter being present. Space requires matter just as matter requires space. These properties of the world were axiomatic to the Cartesian system and ruled the vacuum out of court *ab initio*. Unfortunately, Pascal's meetings with Descartes to discuss the significance of his spectacular mountain-top experiments on air pressure did not go well. Descartes maintained that he had actually proposed that experiments of this sort be performed, but refused to admit that they established the existence of a real physical vacuum, as Pascal always claimed. Pascal did not create a good impression, because after his visit Descartes wrote to Huygens in Holland that he had found Pascal to have 'too much vacuum in his head'!

Pascal ended up entering a public debate, conducted in print, with Descartes' Jesuit tutor Père Noël. Noël sought to defend the non-existence of the vacuum on the ground laid out by his pupil, but also on the ground that the all-pervasive sovereignty of God prevented a vacuum forming anywhere, for this would require an abnegation of the Almighty's power. Noël attacked Pascal's interpretation of his experiment, making a fine linguistic distinction between a vacuum and 'empty space' when it came to evaluating the content of the mercury tube, denying that the space in the tube was the same vacuum that Aristotle had denied could exist:

> "But is this void not the 'interval' of those ancient philosophers that Aristotle attempted to refute . . . or rather the immensity of God that cannot be denied, since God is everywhere? In truth, if this true vacuum is nothing other than the immensity of God, I cannot deny its existence; but likewise one cannot say that this immensity, being nothing but God Himself, a very simple spirit, has parts one separate from the other, which is the definition I give to body, and not that you attribute to my authors, taken from the composition of matter and form."[25]

Pascal did not rise to the bait being dangled here to tempt him into a theological debate about the nature of God with a member of the Jesuits which would have resulted in him being tarred with the same brush as the atheistic atomists like Democritus (the 'ancient philosophers' that Noël mentions). Reclaiming the moral high ground, his reply to Noël cleverly sidestepped the problem with Jesuitical skill:

> "Mysteries concerning the Deity are too holy to be profaned by our disputes; we ought to make them the object of our adoration, not the subject of our discussions: so much so that, without discussing them at all, I submit entirely to whatever those persons decide who have the right to do so."[26]

As other commentators began to see the close connection between Pascal's experiments and the ancient questions concerning the vacuum and the void, Pascal's writings started to stress the 'equilibrium' and 'balance' that the experiments displayed rather than the emptiness. But he was circumspect in his views. His unpublished papers show that he held much firmer opinions than he voiced at the time. In his private writings we find him asking himself about the sense of the Aristotelian abhorrence of the vacuum:

> "Does Nature abhor the vacuum more on top of a mountain than in a valley, and even more so in wet weather than in sunshine?"

Despite the down-to-earth nature of Pascal's study of air pressure, his results had deep and (for some) disturbing implications. His theory of air pressure explained why the height of Torricelli's mercury column should fall as the experiment was carried to high altitude: only the weight of air above the experiment is exerting pressure on the surface of the mercury in the bowl. Suppose that we kept on going – the change in the mercury level has so far been finite, does it not imply that the atmosphere may be finite in mass, surrounding the Earth like a hollowed-out sphere? This would mean that there was ultimately a vacuum out in space, surrounding and enclosing us. Noël argued that it led us to the dangerous conclusion that if this useless vacuum existed in outer space beyond us then it would mean that some of God's creation was of no use. Yet Pascal's arguments won the day. Not until the last half of the twentieth century would it be appreciated how the vastness of the Universe is necessary for the existence of life on a single planet within it.[27]

HOW MUCH OF SPACE IS SPACE?

> "In the United States there is more space where nobody is than where anybody is. That is what makes America what it is."
>
> Gertrude Stein[28]

Fred Hoyle once said that 'space isn't remote at all. It's only an hour's drive away if your car could go straight upwards.'[29] The work begun by Torricelli with such mundane equipment culminated in the discovery that the Earth is surrounded by a gaseous atmosphere that becomes increasingly dilute the further we go from the Earth's surface. Pascal was drawn to speculate what this might ultimately mean for the nature of outer space beyond. Was a true vacuum encircling us or was there simply a medium that grew sparser and sparser beyond the Sun and the planets? In Pascal's time it was not possible to appreciate the enormity of this problem. Today's picture of the Universe allows us to discern the nature of outer space in considerable detail. What we have found is doubly surprising. Matter is organised into a hierarchy of systems of increasing size and decreasing average density. In ascending order of size, there are planets, groups and clusters of stars, and systems of hundreds of billions of stars which come together to form galaxies like our own Milky Way galaxy; then we find galaxies gathered together into clusters that can contain thousands of members and these clusters can be found gravitating together loosely in vast superclusters. In between these regions of greater than average density in the Universe, gas molecules and specks of dust are to be found. The average density of a planet or a star like the Sun is close to one gram per cubic centimetre, which means about 10^{24} atoms per cubic centimetre. This is roughly the density of things we encounter around us. This is vastly greater than the average density of the Universe. If we were to smooth out all the luminous material in the visible universe then we would find only about one atom in every cubic metre of space. This is a far better vacuum than we can make in any terrestrial laboratory by artificial means. There are about one hundred billion galaxies within this visible universe[30] and the average density of material within a galaxy is about one million times greater than that in the visible universe as a whole, and corresponds to about one atom in every cubic centimetre.

Counting up the visible matter in the Universe is only part of the accounting that needs to be done if we are to have a complete inventory of the contents of space. Some matter reveals itself by its luminosity but *all* matter reveals itself by its gravity. When astronomers study the motions of stars in galaxies and galaxies in clusters they find a similar story. The speeds of the moving stars and galaxies are too great for the galaxies and clusters to remain locked together by gravitational attractions between their constituents unless there is about ten times more matter present in some dark unseen form. This is not entirely unexpected. We know that the formation of stars will not be a perfectly efficient process. There will be lots of material that does not get swept up into regions that become dense enough to create the conditions needed to initiate nuclear reactions and start shining. The major mystery is what form this matter takes. It is known to astronomers as the 'dark matter problem'. The obvious first idea that the dark material is just like other matter – atoms, molecules, dust, rocks, planets or very faint stars – does not seem to work. There is a powerful limit on how much material of that sort – luminous or non-luminous – there can be in the Universe in order that the nuclear reactions that produce the lightest elements of helium, deuterium and lithium in the early stages of the Universe give the observed abundances. So we are forced to accept that the dark material that dominates the content of outer space must be another form of matter entirely. The favourite candidate is a population of neutrino-like particles (called WIMPS = weakly interacting massive particles), heavier than ordinary protons and more numerous.[31] They do not take part in nuclear reactions so they avoid the limit on their abundance imposed by the behaviour of nuclear reactions in the early stages of the Universe's history. Such particles are suspected to exist as part of the complement of elementary particles of matter but they would not have been visible in particle physics experiments so far. The theory of the expanding Universe allows the abundance of these particles to be calculated exactly in terms of their mass. If such hypothetical particles do supply the dark matter needed to hold galaxies and clusters together, then we will soon know. They will be detectable in a few years' time in deep underground experiments devised to catch them as they fly through the Earth. A few detections should be made each day in each kilogram of specially designed detection material.

Atoms and molecules, and even neutrino-like particles, are far from all there is pervading outer space. Radiation exists in all wavelengths. The most pervasive and the most significant contributor to the total energy density of the Universe is the sea of microwave photons left over from the hot early stages of the Universe. As the Universe has expanded, these photons have lost energy, increased in wavelength and cooled to a temperature only 2.7 degrees above absolute zero. There are about 411 of these photons in every cubic centimetre of space. That is, there are roughly one billion of these photons for every atom in the Universe.

Our detailed probing of the distribution of matter and radiation in the Universe shows that, as we survey larger and larger volumes of the Universe, the density of material that we find keeps falling until we get out beyond the dimensions of clusters of galaxies (see Figure 3.7). When we reach that scale the clustering of matter starts to fade away and looks more and more like a tiny random perturbation on a smooth sea of matter with a

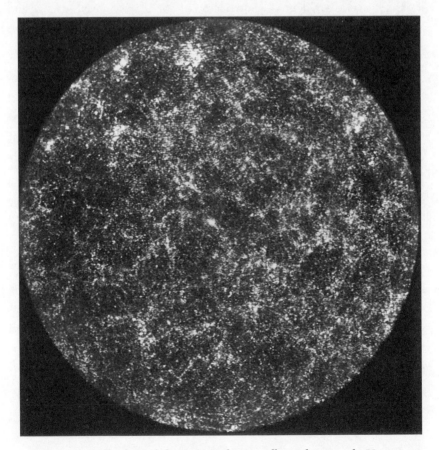

Figure 3.7 *The observed clustering of about a million galaxies in the Universe.*

density of about one atom in every cubic metre. As we look out to the largest visible dimensions of the Universe we find that the deviations from perfect smoothness of the matter and radiation remain at a low level of just one part in one hundred thousand. This shows us that the Universe is not what has become known as a fractal, with the clustering of matter on every scale looking like a magnified image of that on the next larger scale. The clustering of matter appears to peter out before we reach the limit of our telescopes. This is a reflection of the fact that these large aggregates of matter take time to assemble under the influence of gravitational attractions. There is only a finite time available for this process and so its extent is limited.

The Universe appears to be a system of very low density wherever we look. This is no accident. The expansion of the Universe weds its size and age to the gravitational pull of the material that it contains. In order that a universe expands for long enough to allow the building blocks of life to form in the interiors of stars, by a sequence of nuclear reactions, it must be billions of years old. This means that it must be billions of light years in extent and possess a very small average density of matter and a very low temperature. The low temperature and energy of its material ensures that the sky is dark at night. Turn off our local Sun and there is just too little light around in the Universe to brighten the sky. The night is dark, interspersed only by pinpricks of starlight. Universes that contain life must be big and old, dark and cold. If our Universe was less of a vacuum it could not be an abode for living complexity.

In showing what the state of space is today we have rushed ahead to the present. But the vector from Pascal to the Big Bang was not so short. In the next chapter we begin to see what happened to the vacuum in between, how it was transmogrified, banished, restored and ultimately transfigured. We shall find that the concept of the vacuum and the search for evidence for its existence continued to play the same central role in science and philosophy in the nineteenth and twentieth centuries as it did at earlier times.

chapter four

The Drift Towards the Ether

"The idea of an omnipresent medium has considerable attractions for the scientist. It enables him, for example, to explain how such familiar phenomena as light, heat, sound and magnetism can operate over great distances and travel through a seemingly empty space."

Derek Gjertsen[1]

NEWTON AND THE ETHER: TO BE OR NOT TO BE?

"Nothing is enough for the man to whom enough is too little."

Epicurus

Newton's studies of motion and gravity in the second half of the seventeenth century followed a trajectory that would lead to staggering success. He could explain the motions of the Moon and the planets, the shape of the Earth, the tides, the paths of projectiles, the variation of gravity with altitude and depth below the Earth, the motion of bodies when resisted by air pressure, and much else besides. Newton did this by making a spectacular imaginative leap. He formulated laws of motion in terms of an idealised situation. His first law of motion states that 'Bodies acted upon by no forces remain at rest or in motion at constant velocity.' No one had ever seen (or will ever see) a body acted upon by *no* forces, but Newton saw that such an idea provided the benchmark against which one could reliably gauge what was seen. Whilst others had thought that bodies acted upon by no forces just slowed down and stopped, Newton identified all the forces that were acting in any given situation, and thought otherwise. When no motion occurred it was because different forces were in balance, leaving zero net force acting on the body.

Despite the power and simplicity of Newton's ideas, there was an awkward assumption at their heart. Newton had to suppose that there existed something that he called 'absolute space', a sort of fixed background stage in the Universe upon which all the observed motions that his laws governed were played out. Newton's famous laws of motion applied only to motions that were not accelerating relative to this imaginary arena of absolute space.[2] Today, we might approximate it by mapping out an imaginary scaffolding using the most distant, most slowly changing astronomical objects that we can see, the quasars.

Absolute space was a tricky notion. It was the linchpin of Newton's theory but you couldn't observe it, you couldn't feel it and you couldn't do anything to it. It begins to sound as mysterious and elusive as the vacuum itself. It carried with it the added difficulty of not explaining how gravity or light could be propagated through it. One answer to this riddle was to give up the notion that space was empty in between the solid objects dotted around within it and instead imagine that the 'empty' space between contained an extremely dilute fluid that filled its every nook and cranny like a uniform motionless sea. This fluid begins to look like a candidate to replace the entirely mathematical concept of 'absolute space' because motion can always be described as taking place relative to the tenuous fluid.

This great unchanging sea filling all of space became known as the *ether*. It is reminiscent of the elastic substance, or *pneuma*, that the ancient Stoic philosophers proposed as a space filler, which played an active role in their attempts to understand the world. The spreading of sound outwards from its source was interpreted as a motion through the pneuma, like a wave through water. The familiarity of this analogy did much to encourage its adoption as a model for the permeability of all space. Its removal of the need to worry about real vacua ever again was an added attraction.

Newton never displayed any great enthusiasm for the idea and adopted it with some reluctance for want of something more compelling. He recognised that the ether provided a convenient vehicle to understand some of the properties of light and the propagation of its effects through space, but the presence of a fluid would play havoc with the motions of the Moon and the planets. He understood the motions of bodies in liquids and other resisting media very thoroughly and protested that the presence of an all-pervading resisting medium would just retard the motions of the celestial bodies. Eventually, they would grind to a halt.

Torricelli, Pascal and Boyle had investigated a number of properties of the local vacua they could apparently create in their mercury columns. They had shone light through them, and so deduced that light could penetrate the evacuated space; magnetic attraction was not inhibited either; radiative heat passed unimpeded through the jars of empty space; and bodies fell to the Earth under gravity just as they did in air.[3] Newton was well aware of these features of 'empty' space and so wondered if perhaps it was not so empty after all, so that the heat and light could be propagated by the 'vibrations of a much subtler medium than air, which after the air was drawn out remained in the vacuum'.[4]

In trying to sustain this idea, Newton got himself in a very complicated tangle. His first thought was to view light as a stream of minute particles (which we would now call 'photons') that bounced off reflecting surfaces and behaved like the tiniest of perfectly elastic billiard balls. Unfortunately, both he and the Dutch physicist, Christiaan Huygens, had discovered that under some circumstances light did not behave like a stream of little billiard balls at all. Two light beams slightly out of phase with one another could be made to interfere and produce an alternation of dark and light bands. This behaviour is characteristic of waves but not of particles. It can be explained by adding two waves so that the peaks of one wave match the troughs of the other. Newton had observed more colourful consequences of the wavelike behaviour of light, like those we see in the colours created when light passes through oil on the surface of water or scatters off a peacock's tail.

The most useful guide to the issue was the behaviour of sound. Sound is propagated from one point to another by means of undulations in the intervening medium. When we shout across the room it is the vibrations of the molecules of air that carry the energy that we call sound from one place to another. This picture was one that physicists focused upon when thinking about how light moved through empty space. Unfortunately, unlike heat and light, sound was *not* something that was transmitted through the jars of vacuum that Boyle and others had been producing. The extraction of the air from the vacuum tube removed the very medium whose vibration could convey its effects to distant places. Although we *see* the Sun and feel the heat that it radiates to us through the intervening 'empty' space we can't *hear* anything that happens on the surface of the Sun, despite the fantastic violence of those events.

Newton's first attempt to draw these two properties of light together was to imagine that bullets of light must create waves by hitting the ether, just as throwing a stone into water creates a train of waves moving outwards from the impact point. The light would be able to set up an undulatory motion in the ether fluid. Gravity would accelerate them until the accelerating force became equal to the resisting force of the ether and then they would move with constant speed. However, light moves so rapidly that the accelerating force would need to be unrealistically large in order to accelerate the light particles up to 186,000 miles per second so quickly.

Newton was never fully persuaded of the cogency of this ethereal picture and continued to pose questions about the propagation of light and gravity through space without ever convincingly answering them. Newton would not allow himself to lapse into the ancient delusion that some innate property of things called 'gravity' was responsible for the distant action of one mass on another (for this would explain nothing). In his famous correspondence with Richard Bentley[5] about the ways in which his work on gravity and motion could lend support to a new form of Design Argument for the existence of God based upon the precision and invariance of the laws of Nature themselves, rather than the fortuitous outworkings of those laws, Newton revealed his puzzlement at the way gravity could apparently act through a vacuum:[6]

> "It is inconceivable, that inanimate brute matter should, without the mediation of something else, which is not material, operate upon, and affect other matter without mutual contact; as it must do if gravitation, in the sense of Epicurus be essential and inherent in it . . . That gravity should be innate, inherent, and essential to matter, so that one body may act upon another at a distance through a *vacuum*, without the mediation of anything else, by and through which their action and force may be conveyed from one to another, is to me so great an absurdity, that I believe no man who has in philosophical matters a competent faculty of thinking, can ever fall into it. Gravity must be caused by an agent acting constantly according to certain laws; but whether this agent be material or immaterial, I have left to the consideration of my readers."

One can imagine how problematic Newton's picture of forces acting instantaneously at a distance must have been for many of his contemporaries to accept. The rival theory of planetary motion in Newton's time was the vortex theory of Descartes. It viewed the Universe as a great whirlpool of swirling particles whose actions upon one another were conveyed by physical contact (Figure 4.1). Descartes denied that a vacuum existed in space and filled it with a transparent fluid, *matière subtile*, which became a key part of the Cartesian world view.

This picturesque swirling image of the Universe had far more popular appeal than Newton's austere mathematical clockwork. Everyone had seen eddies of turbulent water. The analogy was familiar and convincing: stirring water in one part of the bath tub would propagate effects across the surface to other parts of the water. Descartes appeared to offer a plausible mechanism whereby the effects of gravity could be communicated through space. Yet Descartes' theory failed. It could not explain the

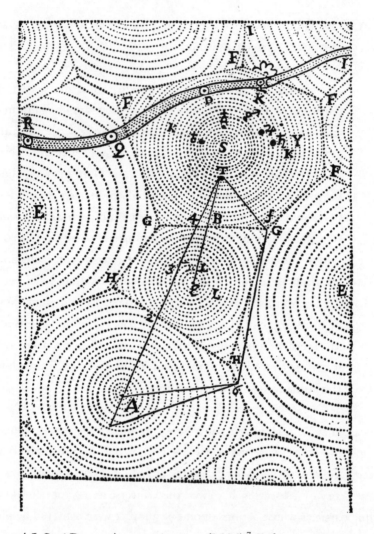

Figure 4.1 *René Descartes' system of vortices (1636).*[7] *Each vortex represents a solar system in a never-ending expanse of solar systems. The centres of the vortices (at the points marked S, E and A) are stars that are shining because of the turbulent motions of the vortices. The sinuous tube passing across the top of the picture is a comet that is moving too fast to be captured by any of the solar systems.*

observed motions of the planets, enshrined in Kepler's famous 'laws'. It was a lesson on the difference between human conceptions of what looks 'natural' and what *is* natural.[8]

The ebb and flow of Newton's views about the ether make interesting reading. In the 1670s he was seeking to persuade Boyle that there existed a subtle ethereal spirit of air (*aere*) because in a vacuum a swinging pendulum continued oscillating for so little longer than it did in air. Newton argued that there must exist another fluid, playing a similar role to that of air, which slows the pendulum even if it is placed in one of Boyle's vacua. He also claimed that some metals could fuse and become heavier even when sealed within a glass container. This, he suggested, meant that some sparse fluid must be passing through the pores in the glass container in order to increase the mass of the metal. A little later he tried to use the ether as a device to explain the reflection and refraction of light, and to persuade Boyle that the non-uniformity of an ether could explain the existence of gravity.

By the 1680s Newton had lost his enthusiasm for the ether. In the *Principia* (1687), he excludes the existence of such a medium permeating masses because it would have an incalculable disturbing influence upon the motion of celestial bodies. Then, in the second book of the *Principia*, he considers directly 'the opinion of some that there is a certain ethereal medium extremely rare and subtle, which freely pervades the pores of all bodies' and seeks to find experiments which might test the idea. He returns to its effects on the swinging pendulum, now interpreting the evidence as indicating that there is no discernible difference in the damping of the pendulum's motion in air or vacuum. Thus he concludes that if an ether exists, its effects must be so subtle as to be indiscernible and so it can safely be ignored for the purposes of explaining gravity and other observable phenomena – a complete about-turn.

Six years later Newton was trying to convince Bentley of the impossibility of an influence like gravity acting instantaneously over great distances, whilst writing to Leibniz that a fine form of matter did indeed fill the heavens above. By the time the second edition of the *Principia* appeared in 1713, Newton had added to the text of the first edition that there was indeed a 'subtle spirit which pervades and lies hid in all gross bodies' and it was this that allowed him to understand the forces of Nature: gravity, heat, light and sound. In what way it did so was not revealed, because it 'cannot be explained in a few words'.

Newton's last views on the ether appear in some of the questions posed at the end of the second edition of his *Opticks* (1717). Here, he claims fresh experimental evidence for the ether's existence by comparing the behaviour of thermometers in air with those sealed in an evacuated tube.[9] Again, the lack of discernible differences in their response to heat convinced Newton that a medium 'more rare and subtle than air' must still be present in the evacuated container in order to transmit the heat from outside. Harking back to his first speculations about the existence of an ether, he then suggests that this subtle medium must be much sparser within dense bodies like the Sun and the planets than it is in the interplanetary space between them. Thus gravity arises because bodies attempt to move from where the ether is denser to where it is sparser[10] – 'every Body endeavouring to go from the denser parts of the Medium towards the rarer' – seeking to even out the distribution.[11] Finally, Newton tried to offer some mechanical explanation for the elusivity of the ether. It was made of particles that are 'exceedingly small' and its elasticity arose from the fact that these particles repel one another. The forces are stronger in small bodies than in large ones in proportion to their mass. The result is[12]

> ". . . a Medium exceedingly more rare and elastick than Air, and by consequence exceedingly less able to resist the motion of Projectiles, and exceedingly more able to press upon gross Bodies, by endeavouring to expand itself."

Newton's speculations on the links between the elusivity and the elasticity of the ether ended with these questions. He never published a detailed theory of the quantitative properties of the ether and its role in mediating the force of gravity. The clarity of his predictions of the effects of gravity and motion stood in contrast to his continual attempts to come to a satisfactory conclusion about the true nature of the vacuum and the way in which forces traverse it. Newton was ahead of his time in almost everything he deduced about the workings of the world, but in the matters of the ether and the vacuum, the jump into the future was too far even for him.

DARKNESS IN THE ETHER

"I have not had a moment's peace or happiness in respect to electromagnetic theory since November 28, 1846. All this time I have been liable to fits of ether dipsomania, kept away at intervals only by rigorous abstention from thought on the subject."

Lord Kelvin[13]

The problem of empty space was entwined with another long-standing riddle: the darkness of the night sky. Descartes' philosophy had rested firmly upon a belief in the impossibility of empty space. He believed in a universe of unending extent. Only matter could have spatial extent and so where there was no matter there could be no space. Everything was moved by forces arising out of direct physical contact. There was no spooky action at a distance across the vacuum. His picture of celestial vortex motions which permitted interactions to occur only by contact (see Figure 4.1) led him to refute the atomists' picture of 'atoms' of matter separated by void. Matter must be continuous and free from voids or other discontinuities. If atoms were introduced into his theory then they would necessarily be in continuous contact with one another and so necessarily extended rather than isolated points of matter as the atomists had imagined.

Descartes' Newtonian opponents rejected his conception of matter moved by purely mechanical laws. Many assumed that the darkness of the night sky between the stars was direct evidence of the infinite and eternal extracosmic void which the ancients maintained existed beyond the edge of a material world of finite size and age. We were seeing through the finite celestial world into the dark void beyond. Thus we see that the Cartesians combined Aristotelian and Epicurean ideas: like Aristotle, they rejected both the vacuum as a physical reality and the atomic nature of matter, but like Epicurus they believed that space had no limit. The Newtonians, by contrast, merged Stoic and Epicurean philosophies: like the Stoics they rejected the idea that the stars were infinite in number and extent, but like the Epicureans they accepted the existence of the vacuum and the basic atomic structure of matter. Later, the Newtonian picture would dispense with its Stoic aspect and use only the Epicurean picture of the boundless population of stars, pictured in Figure 4.2.

Anyone who believed that the Universe contained an infinite distribution of stars was faced with explaining the darkness of the night sky.[14] If one looked out into such an infinite array of stars then it would be like looking into a never-ending forest: one's line of sight would always end on a tree. We should see the entire sky as if it were a single bright starry surface. Evidently, this was not the case.

The hypothesis that space was filled with a tenuous ether created new possibilities for explaining the darkness of the night sky. In the nineteenth

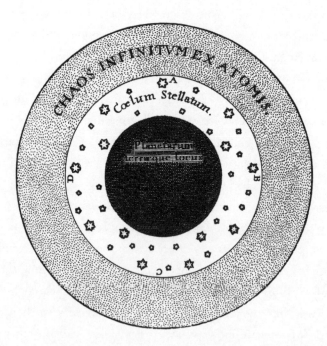

Figure 4.2 *Isaac Newton's view of the Universe in 1667, during his early years in Cambridge.*[15] *This picture combines ancient Epicurean and Stoic conceptions of the cosmos.*

century, the Irish astronomer John Gore suggested[16] that the darkness of interstellar space might be evidence for regions of total vacuum devoid of both matter and ether:

> "It has been argued by some astronomers that the number of the stars must be limited, or on the supposition of an infinite number uniformly scattered through space, it would follow that the whole heavens should shine with a uniform light, probably equal to that of the sun."[17]

Gore and the Canadian astronomer Simon Newcomb[18] both believed that the puzzle of the dark night sky would be solved if our Milky Way galaxy were shielded from the stars and nebulae beyond by a perfect vacuum region across which starlight could not travel. Thermodynamically, this sounds rather odd. What happens to the starlight when it impinges upon this impervious vacuum region? They suggested that it was reflected back so that

> "we may consider . . . the reflecting vacuum as forming the internal surface of a hollow sphere."

In their scenario each galaxy of stars and ordinary material is surrounded by a spherical 'halo' of ether, but the intergalactic region between the ether halos is a perfect vacuum which light cannot penetrate (see Figure 4.3).

One can see that for all practical purposes, the other galaxies, with their ethereal halos, might as well not exist. They are unobservable in principle. The darkness of the night sky is really being explained by supposing that the Universe is astronomically finite and contains very few stars. All the rest are an optical illusion. Unfortunately, this does not work. If each galaxy is surrounded by a mirror of perfect vacuum then the starlight from the stars it contains will be bounced back and forth across the galaxy and end up contributing a similar amount of light to the visible sky as would be incoming from the other galaxies.

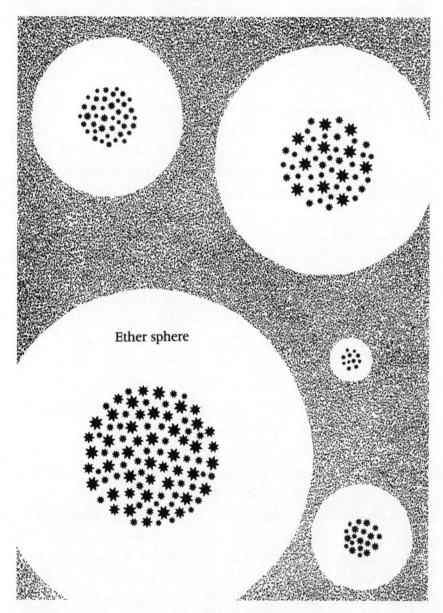

Ether sphere

Figure 4.3 *Newcomb and Gore's solution to the puzzle of the darkness of the night sky.*[19] *Each galaxy of stars is surrounded by a sphere of ether. The space between the galaxies contains no ether and so cannot transmit light. The spheres of ether act as if they are reflecting mirrors and prevent observers within them receiving light from other spheres.*

NATURAL THEOLOGY OF THE ETHER

"If God had meant us to do philosophy he would have created us."

Marek Kohn[20]

During the eighteenth and nineteenth centuries theologians were greatly impressed by arguments for the existence of God which cast Him in the role of cosmic Designer. The existence of such a Designer, they argued, was evident from the structure of the world around us. This structure had two telling strands. First, there were the apparent contrivances of the living world. Animals seemed to inhabit environments that were tailor-made for their needs. What more perfect design could be found than the camouflage markings on the coat of an animal that merged so exactly with its surroundings? These arguments about the ways in which the outcomes of Nature's laws are in mutual harmony had been supplemented by the more sophisticated Design Argument based upon the success of Newton's elucidation of simple laws of Nature that Richard Bentley promulgated. This second version of the Design Argument pointed to the simplicity and mathematical power of those simple, all-encompassing, Newtonian laws as the primary evidence for a Cosmic Lawgiver who framed them.[21]

In all of these natural theological discussions the Universe was viewed as a harmonious whole in which all its components were wisely and optimally integrated into a grand cosmic scheme. Humanity was a beneficiary of this scheme, but only the most naive pictures of Design would insist on making humanity's well-being the goal or final cause of the whole creation. The ether fitted into this teleological conception of the Universe because it resolved the old objection that a void space serves no purpose and implies that the Deity was responsible for making things that were a waste of space. The ether plugged this critical gap by doing away with the purposeless emptiness. It found itself elevated in the minds of some the-

ologians to play a role just a little lower than the angels as the main second-
ary cause by which God regulated the motions of the celestial bodies. For
example, John Cook argues that

> "Ether is the Rudder of the Universe, or as the Rod, or
> whatever you will liken it to, in the Hand of the Almighty,
> by which he naturally rules and governs all material cre-
> ated Beings . . . Now how beautiful is this Contrivance in
> God."[22]

This line of argument was developed most elaborately by William
Whewell in his contribution to the famous Bridgewater Treatises on
Natural Theology, a series of works by distinguished nineteenth-century
scholars seeking to provide support for Christian religious belief by appeal
to scientific discovery. Whewell's volume[23] dealt with the contribution of
astronomy and physics to the argument. Since he was a strong supporter of
Huygens' wave theory of light, the ether played a key role in his conception
of the physical universe and he was greatly persuaded of its crucial role in
the theological scheme of things as well. He argued that ether was provi-
dentially designed by the Almighty in order to enable us to see the Universe
with our visible sense. It was one of the three fundamental substances in
the Universe, beside matter and fluid.[24] Without it, the Universe would be
dead, inert and unknowable. Its very existence was thus evidence for the
wisdom, goodness and anthropocentric good intentions of God.

Amongst notable scientists, the most speculative views on the ether
are to be found in the works of the Scottish physicist Peter Guthrie Tait,
who is famous for his joint work with Lord Kelvin and his pioneering ideas
in the mathematical theory of knots. In 1875, Tait co-authored a popular
science book with Balfour Stewart which bore the title *The Unseen universe;
or, physical speculations on a future state.*[25] Its purpose was to demonstrate the
harmony of religion and science and, in seeking to do this, it had some
remarkable things to say about the ether.

Stewart and Tait suggested that all matter was composed of particles
of ether, but these ether particles were composed of an even subtler collec-

tion of ether particles, and so on, ad infinitum. This hierarchy of ethers was arranged in an ascending one-way street of energies, so that lower-order ethers could always form from a higher, but not vice versa. Stewart and Tait imagined their staircase of ethers rising, like Jacob's ladder, to attain infinite energy and ultimately becoming eternal and co-equal with God. The creation of the world was simply the cascade of energy down the spectrum of ethers so that it became localised in matter at the lowest levels, those we see around us and in which we have our being.

A DECISIVE EXPERIMENT

"Now the sirens have a still more fatal weapon than their song, namely their silence . . . someone might possibly have escaped from their singing; but from their silence never."

Franz Kafka[26]

In the middle of the nineteenth century, it was the accepted view of almost all scientists that space was filled with a ubiquitous ethereal fluid. There was no vacuum. All forces and interactions were mediated by the presence of ether, either by waves of ether or by vortices. The favoured scenario was one in which the ether was, on average, stationary; others suggested that it was dragged around by the daily rotation of the Earth and by its annual orbit of the Sun. To question this picture seemed rather foolish, a little like questioning whether the Earth possessed an atmosphere of air. The existence of the ether was fast becoming one of those scientific truths we hold to be self-evident. Yet, whilst its existence was not doubted, its physical characteristics were the subject of lively debate.[27] Some held it to be thin and tenuous, others argued it was an elastic solid, and others still that its properties changed according to the ambient conditions. In such a confused atmosphere speculative theories abound, and all manner of contrived additional properties are easily invented to modify the favoured hypothesis in the face of new objections or awkward facts. What is needed is a decisive experiment. Just as Torricelli cut through the convoluted debates about the possibility of the physical vacuum by providing an experimental window into the question, so it would be with the ether. The impetus was to come from a side of the Atlantic where few would have expected the next great step in our understanding of motion to be taken.

Albert Michelson was born on 19 December 1852 in the small town of Strelno near the Polish-German border.[28] Technically, Strelno had been in Germany since the time of Frederick the Great but its traditions were Polish, like its citizens, and it was less than eighty miles from Copernicus' birthplace. In the face of political upheaval and persecution, the Michelson family joined thousands of other Polish emigrants to the United States when Albert was just two years old. After working as a jeweller for a time in New York, Albert's father Samuel joined the gold rush to California to seek his fortune. Soon afterwards, California became a State of the Union and grew rather prosperous. Samuel Michelson prospered as well and set up his own store in Calaveras County. The rest of the family eventually joined him after a formidable sea voyage to Panama followed by a dangerous overland trek across the neck of the continent (in the days before the canal) to the Pacific, where they took another boat to San Francisco before the final overland journey to the Gold Towns. There, in a wild-west frontier atmosphere, far from the world of learning and traditional culture, the young Michelson spent his early formative years. As a child, he was exceptionally gifted at constructing mechanical devices and showed an early aptitude for mathematics, together with a fascination for the rocks and minerals that the miners dug out of the ground. On reaching his thirteenth birthday, he was sent away to high school in San Francisco, and after graduating successfully three years later he entered a fierce competition for a place at the US Naval Academy in Annapolis, Maryland. Alas, he didn't get the place. He tied in the examinations with a younger candidate from a poor background who was given the casting vote by the selection board despite the mountain of letters written in support of Michelson.

Michelson didn't give up. Such was his determination to be admitted to the Academy that he appealed directly to President Grant for a further place to be created. Learning of the President's daily routine of walking his dog, he travelled to Washington and waited on the White House steps for

him to return. Grant listened patiently to the teenager's request but said
there was really nothing he could do. All the places at the college were
filled. But then he remembered a letter he had received from Michelson's
congressman arguing his case on the ground of his father's great contribution commercially and politically to the Republican cause: to reward young
Michelson would bring further support to the President in his home state.
For whatever reason, the President decided to intervene and sent the young
Michelson to see the Superintendent of the Naval Academy in person.
Interviews followed and after just a few days Michelson heard that an extra
place had been created at the Academy for new entrants that year and it had
been awarded to him. He entered as a cadet midshipman and gradually distinguished himself at the college in all the science courses, less so in military matters.[29] After graduating, and spending a short spell at sea, he was
appointed instructor in physics and chemistry at the Academy and began to
develop his expertise in optics and experimental physics. His first distinguished contribution to science was a precision measurement of the speed
of light. After this work was completed, in 1880, Michelson took a period
of leave from the Navy and took his family to Europe. It was a trip that was
to change the direction of science.

Michelson spent two years moving between some of the leading
European universities, learning about the new developments in physics
and, inevitably, listening to some of the foremost theoretical physicists
expound their theories of the ether – the great puzzle of the day. It became
a source of continuing fascination for Michelson. Did this strangely elusive
medium exist or not? Was there a way of measuring it?

James Clerk Maxwell had suggested[30] that by checking whether the
speed of light was the same in different directions we would learn something about the motion of the ether stream through which the light had to
propagate; for

> "If it were possible to determine the velocity of light by
> observing the time it takes to travel between one station
> and another on the earth's surface, we might by comparing
> the observed velocity in opposite directions determine

the velocity of the ether with respect to these terrestrial stations."

But Maxwell doubted whether it would be possible to conduct this experiment and discover the answer. Michelson ignored these pessimistic predictions. He saw that there was a straightforward way to realise Maxwell's inspired suggestion. Suppose that the ether is not moving, so it specifies some state of absolute rest. Then we must be moving through it, as the Earth spins on its axis and orbits the Sun. A suitable detector might be able to measure the wind of ether that our movement through it would create, just like a cyclist feels the wind on his face as he rides through still air. If the ether was moving, then we should feel different effects when we move upstream and downstream within it.

Michelson began to develop his ideas by drawing up simple analogies. Moving through the ether should be like swimming in a river. The flow of the river is the flow of ether past us caused by the Earth's motion through the stationary sea of ether. Now imagine a swimmer who makes two return trips in the river. The first is across the river at right angles to the flow; the second is downstream and then back upstream. In both cases he ends up at the same point at which he began; the two round-trip paths are shown in Figure 4.4. If the same total distance is swum in each case then it is always

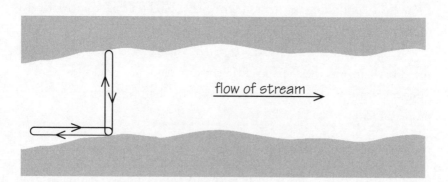

flow of stream

Figure 4.4 *A swimmer makes two round trips of equal distance: one across the river and back, the other upstream and downstream.*

quicker to swim the cross-river circuit than the down-and-upstream circuit. To see this, let's do a simple example. Suppose the river flows at a speed of 0.4 metres per second and the swimmer can swim at a speed of 0.5 metres per second in still water. Each leg to be swum will be 90 metres in length.

The swim downstream followed by the return upstream has the swimmer moving at $0.5 + 0.4 = 0.9$ metres per second relative to the bank on the downward trip. The time to swim 90 metres is therefore $90 \div 0.9 = 100$ seconds. Returning upstream his speed relative to the bank is only $0.5 - 0.4 = 0.1$ metres per second and he takes $90 \div 0.1 = 900$ seconds to make it back to the start. The total round-trip time is therefore $900 + 100 = 1000$ seconds.

Now consider the cross-river route. He will find it just as hard to swim each way as he is always swimming at right angles to the current. His speed perpendicular to the flow of the river is given by an application of Pythagoras' theorem for triangles, applied to the velocities. The actual speed he can swim across the river will be equal to the square root of $0.5^2 - 0.4^2 = 0.09$, which is 0.3 metres per second (see Figure 4.5). Thus he can swim 90 metres in $90 \div 0.3 = 300$ seconds. The total time he takes to swim 90 metres across the river and 90 metres back is therefore 600 seconds. This is different from the round-trip time up and downstream because of the speed of flow of the river. Only if the speed of flow of the river is zero will the two round-trip times be the same.

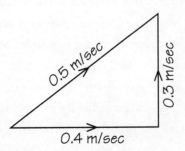

Figure 4.5 *The speed that the swimmer can achieve across the river is given by Pythagoras' theorem applied to the triangle of velocities.*

Michelson concluded that the same should happen if light was 'swimming' through the ether. Two light rays emitted in perpendicular directions and reflected back to their starting point should take different times to complete their round trips over the same distance because, like the swimmer in the river, they would experience different total amounts of drag from the flowing ether. Most important of all, *if there was no ether* then the two round-trip travel times by the different light rays should be *exactly the same*.

Michelson conceived of a beautiful experiment to test the ether hypothesis. He sent two beams of light simultaneously in directions at right angles to each other and then reflected them back along the directions they had come. The ether hypothesis could then be tested by checking if they both returned to the starting point at the same moment. The experimental set-up is shown in Figure 4.6. Very high precision was possible by exploiting the wavelike character of light. If the light waves arrived back at the same point slightly out of phase with the source waves then there would a slight darkening caused by the overlap of peaks of one light wave with troughs of the other and a brightening where peaks overlap peaks and troughs overlap troughs. The phenomenon, known as *interference*, creates an alternating sequence of dark and light bands.[31] In Michelson's experiment, seeing no interference pattern of alternating fringes would mean that there could be no effect of the ether slowing the light in one direction but not in a direction perpendicular to it.[32]

Although the concept of the experiment was simple, the execution was a considerable challenge. The speed of light is 186,000 miles per second whilst the speed of the Earth in its annual orbit around the Sun is only about 18 miles per second. Extraordinary care and accuracy was required if the experimental measurements were going to be accurate and not disrupted by measurement errors and other fluctuations in the experimental set-up. To get some idea of the challenge, if the ether did exist then the tiny difference that should be detected in the light-travel time for rays moving in the two directions would be just one-half of the square of the ratio of the speed of the Earth to the speed of light — less than one part in one hundred million! In order to convince scientists that there was no ether-

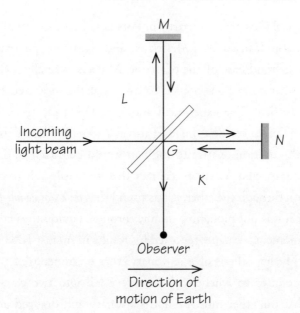

Figure 4.6 *A simplified sketch of Michelson's experiment. A light beam is divided by a partially reflecting glass plate at G into two beams at right angles. One moves along the path of length L, the other along the path of length K. Each is reflected back by mirrors at M and N and the two light beams are recombined at G and observed. If the times required by the two light beams to travel their two paths are different they will be out of phase when they recombine at G and interference bands will be created.*

induced time difference, the accuracy of the measurement had to be better than that.

Fortunately for Michelson, the famous telephone engineer Alexander Graham Bell put up the money to fund the experiment and the interferometer was built in Berlin in 1881. Michelson made his first attempt at the experiment at the University of Berlin, in the laboratory of the famous German physicist Hermann von Helmholtz. He immediately encountered problems. He had to make sure that his array of mirrors was kept at constant temperature by surrounding the entire experiment with melting ice at zero degrees Centigrade, then he had to deal with the vibrations created by the Berlin traffic thundering past outside. The traffic noise ultimately proved unbeatable in Berlin, so he dismantled his apparatus and moved it to the Astrophysical Observatory nearby in Potsdam. Faced now with only the modest vibrations caused by pedestrians, and with his apparatus firmly mounted on the rigid base of the telescope, Michelson finally succeeded in creating the quiet conditions needed to carry out the measurement to the accuracy he needed. The experiment was repeated many times with the apparatus in different orientations, and also at different times of the year, so that the Earth's motion relative to the Sun would be different. The result was entirely unexpected. With an accuracy that was easily able to detect the Earth's motion through the ether, it was found that *there was no interference pattern*. The Earth was not ploughing its way through a ubiquitous ether at all. Michelson's momentous paper reported his results in August 1881 and concluded that 'The hypothesis of a stationary ether is erroneous'.[33]

The responses to Michelson's discovery fell into two camps. Some concluded that the ether must be non-stationary and dragged around by the Earth as it orbits the Sun so that there is no relative motion between the ether and Earth; others simply concluded that the ether didn't exist after all. Michelson remained agnostic about the theoretical interpretation of his result.

Michelson returned to a new position at the Case Institute of Technology in Cleveland.[34] There he gained a new collaborator, an American chemistry professor fifteen years his senior, called Edward Morley. Morley was deeply religious. His original training had been in theology and he only turned to chemistry, a self-taught hobby, when he was unable to enter the ministry. Michelson, by contrast, was a religious agnostic. But what they had in common was great skill and ingenuity with scientific instruments and experimental design. Together they repeated Michelson's experiment to discover if the speed of light was the same in different directions of space. When they finished analysing their results in June 1887 there again were no interference fringes. Light was travelling at the same speed in different directions irrespective of the speed of its source through space. There was no stationary ether.[35] This was an incredible conclusion. It meant that if you fired a light beam from a moving source it would be found to have the same speed relative to the ground that it would have if the source were stationary. Light moved like nothing else that had ever been seen.

THE AMAZING SHRINKING MAN

"A mathematician may say anything he pleases, but a physicist must be at least partially sane."

Josiah Willard Gibbs

How could the ether still exist in the face of the null result of Michelson and Morley? An answer was suggested first in 1889 by George FitzGerald at Trinity College Dublin and then developed independently a little later by the Dutch physicist Hendrik Lorentz at Leiden. They suggested that the length of an object will be seen to diminish if it moves at increasing speeds.[36] If we take two rulers and hold one still on the Earth but let the other fly past at high speed parallel to it then, as the moving ruler passes by, it will be seen to be shorter than the stationary one. This sounded crazy, even to physicists, but FitzGerald and Lorentz derived their claim from the properties of Maxwell's theory of light and electromagnetism. FitzGerald even tried to explain the basis of the contraction by arguing that the inter-molecular forces holding solid bodies together are probably electromagnetic in origin and so were likely to be affected if they moved through the ether. He thought that an increase in their attractiveness could be responsible for drawing molecules closer together and reducing the length of any chain they formed.

The amount of the FitzGerald-Lorentz shrinkage was predicted to be very small. Lengths of moving objects would contract by a factor equal to $\sqrt{(1-v^2/c^2)}$, where v is their speed and c is the speed of light. For a speed of 500 km per hour, we are looking at a contraction that is not much bigger than one hundred billionth of one per cent.

FitzGerald had noticed that if this $\sqrt{(1-v^2/c^2)}$ correction factor was applied to the analysis of Michelson's apparatus fixed on the Earth's surface as it moved around the Sun, it could explain why Michelson measured no effect from the ether. The arm of the interferometer contracts by a factor $\sqrt{(1-v^2/c^2)}$ in the direction of its motion through the ether at a speed v. At an orbital speed of 29 kilometres per second this results in a contraction of only one part in 200,000,000 in the direction of the Earth's orbital motion. The length of the arm perpendicular to the ether's motion is unaffected. This small contraction effect exactly counterbalances the time delay expected from the presence of a stationary ether. If the FitzGerald-Lorentz contraction occurred then it allowed the existence of a stationary ether to be reconciled with the null result of the Michelson-Morley experiment. Space need not be empty after all.

The ideas of FitzGerald and Lorentz[37] were regarded as extremely speculative by most physicists of their day, and not taken very seriously as a defence of the ether. They were considered to be purely mathematical excursions devoid of real physical motivation. Attitudes began to change in 1901 when a young German physicist, Walter Kaufmann, studied the fast-moving electrons, called beta particles, emitted by radioactive elements, and showed that the measured masses of these electrons were also dependent on their speeds, just as Lorentz had predicted. Their masses increased with increasing speed, v, to a value equal to their mass when at rest divided by the FitzGerald-Lorentz factor $\sqrt{(1-v^2/c^2)}$.

The most awkward feature of these attempts to evade casting out the ether was the need to distinguish between a system that was moving and one that was not in some absolute sense. It is all very well to enter a value for v which corresponds to the Earth's orbital velocity around the Sun in the FitzGerald contraction formula, but what if the Sun and its local group of stars are themselves in motion? What velocity do we use for v and with respect to what do we measure it?

EINSTEIN AND THE END OF THE OLD ETHER

"Navy: Please divert your course 15 degrees to the North to avoid a collision.

Civilian: Recommend you divert *your* course 15 degrees to South to avoid a collision.

Navy: This is the Captain of a US Navy ship. I say again, divert *your* course.

Civilian: No, I say again, divert *your* course.

Navy: This is the aircraft carrier *Enterprise*. We are a large warship of the US Navy. *Divert your course now!!*

Civilian: This is a lighthouse. Your call."

Canadian naval radio conversation[38]

The nineteenth century ended with a confusing collection of loose ends dangling from Michelson and Morley's crucial experiment: the absence of the expected ether effect, the need to know the absolute value of velocity, the possibility that motion affects length and mass, and the significance of the speed of light. Albert Einstein made his first appearance on the scientific stage in 1905, at twenty-six years of age, by solving all of these problems at once in an announcement[39] of what has become known as the 'special theory of relativity'. The English translation of his famous paper has the innocuous-sounding title 'On the Electrodynamics of Moving Bodies'.

Einstein abandoned the idea that there was any such thing as absolute motion, absolute space or absolute time. All motion was relative and two postulates, that the laws of motion and those of electromagnetism must be found to be the same by all experimenters moving at constant velocities relative to one another, and that the velocity of light in empty space must be measured to be the same by all observers regardless of their motion, sufficed to explain everything. This enabled him to deduce as a simple consequence the precise laws for length, mass and time change proposed by FitzGerald and Lorentz. This theory reduced to Newton's classical theory of motion when the motions occurred at velocities far less than that of light but behaved in quite different ways as velocities approached that of light in empty space. Newton's theory was seen to be a limiting case of Einstein's.

This feature of a successful new theory of physics is worth dwelling on as it is overlooked by many commentators. Recently, there have been many newspaper polls to pick the most influential thinkers of the millennium. Newton has topped some polls, but finished behind Shakespeare, Einstein and Darwin in others. On one occasion, Newton's lower position was justified on the grounds that some of his laws of motion had been shown to be 'wrong' by the work of Einstein. Indeed, the outsider might be tempted to think that the whole progression of our knowledge about the workings of Nature is replacing wrong theories by new ones which we think are right for a while but which will eventually be found to be wrong as well. Thus, the only sure thing about the currently favoured theory is that it will prove to be as wrong as its predecessors.

This caricature misses the key feature. When an important change takes place in science, in which a new theory takes the stage, the incoming theory is generally an extension of the old theory which has the property of becoming more and more like the old theory in some limiting situation. In effect, it reveals that the old theory was an approximation (usually a very good one) to the new one that holds under a particular range of conditions. Thus, Einstein's special theory of relativity becomes Newton's theory of motion when speeds are far less than that of light, Einstein's general theory of relativity becomes Newton's theory of gravity when gravitational fields are weak and bodies move at speeds less than that of light. In recent years we have even begun to map out what the successor to Einstein's theory may look like. It appears that Einstein's theory of general relativity is a limiting, low-energy case of a far deeper and wider theory, which has been dubbed M theory.[40]

In some respects this pattern of 'limiting' correspondence is to be expected. The old theory has been useful because it has explained a significant body of experimental evidence. This evidence must continue to be well explained by the new theory. So, wherever physics goes in the next millennium, if there are still high-school students learning it in a thousand years' time, they will still be learning Newton's laws of motion. Their application to everyday problems of low-speed motion will never cease. Although they are not the whole truth, they are a wonderful approximation at low speeds[41] to a part of the whole truth. They are not 'wrong' unless you try to apply them to motions close to the speed of light.

Einstein's brilliant success in bringing together all that was known about motion into a simple and mathematically precise theory was the end of the nineteenth-century ether. Einstein's theory had no need of any ether to convey the properties of light and electricity. His postulate that the speed of light must be the same for all observers had the FitzGerald-Lorentz contraction as a direct consequence, and the non-detection of any light-delay effect in the Michelson-Morley experiment was a key prediction of his theory. Many years later, on 15 January 1931, Einstein made a speech in Pasadena to an audience containing many of the world's greatest physicists. Michelson was there, making what would turn out to be his last public appearance before his death four months later. Einstein paid tribute to the importance of the experiment that Michelson first performed in guiding physicists to their revolutionary new picture of space, time and motion:[42]

> "You, my honoured Dr. Michelson, began this work when I was only a little youngster, hardly three feet high. It was you who led the physicists into new paths, and through your marvellous experimental work paved the way for the development of the Theory of Relativity. You uncovered an insidious defect in the ether theory of light, as it then existed, and stimulated the ideas of H.A. Lorentz and FitzGerald, out of which the Special Theory of Relativity developed. Without your work this theory would today be scarcely more than an interesting speculation; it was your verification which first set the theory on a real basis."

In fact, Einstein's career intersected with the ether on many occasions. It only became known after his death that at the age of fifteen he became interested in the stationary elastic ether. He even wrote an article about what happens to the state of the ether when an electric current is turned on, which was not published until 1971.[43] Later, he also contemplated carrying out experiments which would be able to verify the existence of an ether. Gradually, he began to doubt its existence. In 1899, he wrote to his girlfriend Mileva Maric of his doubts:

> "I am more and more convinced that the electrodynamics of the bodies in motion, such as it is presented today, does not correspond with reality and that it will be possible to formulate it in a simpler way. The introduction of the word 'ether' in the theories of electricity leads to the idea of a medium about the motion of which we speak without the possibility, as I think, to attribute any physical sense to such a speech."[44]

As a student he learned about Lorentz's theory of electrodynamics, and the role played by the ether, in his course textbooks. When his thinking drove him towards his new theory of motion, he found he had no need of the ether or of a vacuum with any special properties. It was enough to be able to talk about bodies moving in space and through time. That space was empty unless one chose to add further ingredients to it. It was a matter for investigation whether one needed to include a magnetic or an electric field everywhere in the Universe. If such fields of force were ubiquitous, then his theory could handle them but, equally, it could apply itself to the movements of bodies in a completely empty space.[45]

The developments in our understanding of matter and motion in the first few years of the twentieth century brought to an end what is sometimes called 'classical' physics. Just a few years before there had been serious speculation that the work of physics was all but done. There were refinements to make, further decimal places to establish in experimental accuracy, but all the great physical principles of Nature were thought by some to have been mapped out. The details merely had to be filled in. The discoveries of the quantum theory of matter and the relativity of motion changed everything. New vistas opened up. But they were vistas that did not need a theory of the vacuum, or even a clear notion of what it was.

Emphasis switched to the study of how fields and particles influenced one another. Ancient dilemmas like that of the extracosmic void or the nature of absolute space were issues that philosophers still talked about but they were not subjects that promised new insights. Physicists seemed rather relieved to be able to ignore the vacuum for a change, rather than find it like the proverbial tail wagging the dog, in steering the direction of theories of electricity, magnetism and motion.

This brief era of nothingless physics was soon ended. Within ten years of Einstein's issue of a redundancy notice to the ether, the issue of the vacuum was back in a central and puzzling place in scientific thinking. The deeper and wider extensions of the special theory of relativity and the quantum picture of matter would reinstate the vacuum in a central position from which it has yet to be dislodged at the start of another century.

chapter five

Whatever Happened to Zero?

> "The fool saith in his heart that there is no empty set. But if that were so, then the set of all such sets would be empty, and hence *it* would be the empty set."
>
> Wesley Salmon

ABSOLUTE TRUTH – WHERE IS IT TO BE FOUND?

> "As lines, so loves oblique may well
> Themselves in every angle greet
> But ours, so truly parallel,
> Though infinite can never meet."
>
> Andrew Marvell, 'Definition of Love'[1]

Like the Grand Old Duke of York, who marched his men to the top of the hill and marched them down again, nineteenth-century physicists had been busily filling the ancient void with ether and emptying it out again. In the meantime, what had been happening to zero, that handy little circle that provided the final piece in the jigsaw of symbols that went to make our modern system of arithmetic?

During the nineteenth century, mathematics began to move in a new direction and its scope expanded beyond the paths mapped out by the ancients. For them, mathematics provided a way of making precise statements about quantities, lines, angles and points. It was divided into arithmetic, algebra and geometry, and formed a vital part of the ancient curriculum because it offered something that only theology would also dare to claim – a glimpse into the realm of absolute truth. The most important exemplar was geometry. It was the most impressive and powerful instrument wielded by mathematicians. Euclid created a beautiful framework of axioms and deductions that led to truths called 'theorems'. These truths led to new knowledge of the motions of the planets, new techniques for engineering and art; Newton's greatest insights were achieved by means of geometry.

Geometry was not seen as merely an approximation to the true nature of things, it was part of the absolute truth about the Universe. Like part of some holy writ, the great theorems of Euclid were studied in their original language for thousands of years. They were true, perfectly so, and they provided human beings with a glimpse of absolute truths. God was many things but he was undoubtedly also a geometer.

We begin to see why mathematics was of such importance to theologians and philosophers. With no knowledge of mathematics you might have been persuaded that the search for absolute truth was a hopeless quest. How could we fathom its bottomless complexity given the approximate and incomplete nature of our understanding of everything else in the world around us? How could a theologian claim to know anything about the nature of God or the nature of the Universe in the way that medieval philosophers seemed to do so confidently in their pronouncements about the vacuum and the void? Their justification was in the success of Euclid's geometry. It was the prime example of our success in understanding a part of the ultimate truth of things. And if we could succeed there, why not elsewhere as well? Euclid's geometry was not just a mathematician's game, a rough approximation to things, or a piece of 'pure' mathematics devoid of contact with reality. It was the way the world was. A similar exalted status was afforded the system of logic that Aristotle introduced as the means by which the truth or falsity of deductions made from premises could be ascertained. Aristotle's logic was accepted as being true and perfectly representative of the working of the human mind. It was the one and only way of reasoning infallibly.[2]

Euclid's geometry is a logical system that defines a number of concepts, makes a number of initial assumptions, sets down what rules of reasoning are to be allowed, and then allows an edifice of geometrical truths to be deduced by applying the rules of reasoning to the concepts and axioms. It is rather like a game of chess. There are pieces and rules governing their movement together with a starting position for all the pieces on the board. Applying the rules to the pieces produces a sequence of positions for the pieces on the board. Each possible configuration of pieces that can be reached from the starting position could be regarded as a 'theorem' of chess. Sometimes one encounters inverse chess puzzles that challenge you to decide whether or not a given board position could have been the result of a real game or not.

Euclid's geometry described points, lines and angles on flat surfaces. It is now sometimes called 'plane geometry'. He set out definitions of twenty-three necessary concepts and five postulates. To get the flavour of how pedantically precise Euclid was, and how little he took for granted, here are a selection of his definitions:[3]

Definition 1: A point is that which has no part.

Definition 2: A line is a breadthless length.

Definition 4: A straight line is a line that lies evenly with the points on itself.

Definition 23: Parallel straight lines are straight lines that, being in the same plane and being produced indefinitely in both directions, do not meet one another in either direction.

Euclid's aim was to avoid using pictures or practical experience. All truths of plane geometry must be deduced by using these definitions and five other axioms or 'postulates' from which everything follows by logical reasoning alone. This arena for plane geometry was circumscribed most potently by one of its axioms, the fifth, which stated that parallel lines never meet.[4] Usually it is known as the 'parallel postulate'. There had always been special interest in this axiom because some mathematicians suspected that it might be an unnecessary stipulation: they believed it could be deduced as a logical consequence of Euclid's other axioms. Many claims were made at different times to have proved the parallel postulate from the other axioms, but all were found to have cheated in some way, subtly assuming precisely what was to be proved along the way.

The great success of Euclidean geometry had done more than merely help architects and astronomers. It had established a style of reasoning, wherein truths were deduced by the application of definite rules of reasoning from a collection of self-evident axioms. Theology and philosophy had used this 'axiomatic method', and most forms of philosophical argument followed its general pattern. In extreme cases, as in the works of the Dutch philosopher Spinoza, philosophical propositions were laid out like the definitions, axioms, theorems and proofs to be found in Euclid's works.[5]

This confidence was suddenly undermined. Mathematicians discovered that Euclid's geometry of flat surfaces was not the one and only logically consistent geometry. Carl Friedrich Gauss (1777–1855), Nikolai Lobachevski (1793–1856) and Janos Bolyai (1802–1860) all contributed to the revolutionary idea of giving up the quest to prove Euclid's parallel postulate from his other axioms and, instead, see what happens if one assumes that it is *false*.[6] This revealed that the fifth axiom was by no means a consequence of the other axioms. In fact, it could be replaced by another axiom and the system would still be self-consistent.[7] It would still describe a geometry but not one that exists on a flat surface.

There exist other, non-Euclidean, geometries that describe the logical interrelationships between points and lines on curved surfaces (see Figure 5.1). Such geometries are not merely of academic interest. Indeed, one of them describes the geometry on the Earth's surface over large distances when we assume the Earth to be perfectly spherical. Euclid's geometry of flat surfaces happens to be a very good approximation locally only because the Earth is so large that its curvature will not be noticed when surveying small distances. Thus, a stonemason can use Euclidean geometry, so can a tourist travelling about town, but an ocean-going yachtsman cannot.

This simple mathematical discovery revealed Euclidean geometry to be but one of many possible logically self-consistent systems of geometry. All but one of these possibilities was *non-Euclidean*. None had the status of absolute truth. Each was appropriate for describing measurements on a different type of surface, which may or may not exist in reality. With this, the

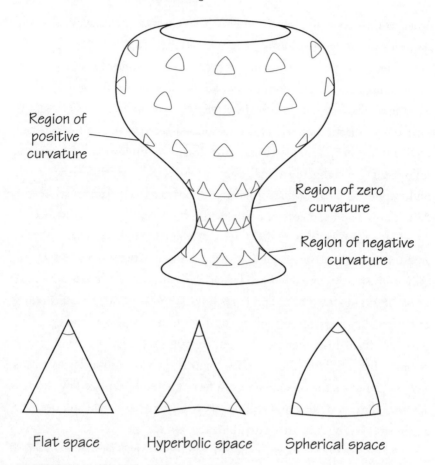

Figure 5.1 *A vase whose surface displays regions of positive, negative and zero curvature. These three geometries are defined by the sum of the interior angles of a triangle formed by the shortest distances between three points. The sum is 180 degrees for a flat 'Euclidean' space, less than 180 degrees for a negatively curved 'hyperbolic' space and more than 180 degrees for a positively curved 'spherical' space.*

philosophical status of Euclidean geometry was undermined. It could no longer be exhibited as an example of our grasp of absolute truth. Mathematical relativism was born.

From this discovery would spring a variety of forms of relativism about our understanding of the world.[8] There would be talk of non-Euclidean models of government, of economics, and of anthropology.

'Non-Euclidean' became a byword for non-absolute knowledge. It also served to illustrate most vividly the gap between mathematics and the natural world. Mathematics was much bigger than physical reality. There were mathematical systems that described aspects of Nature, but there were others that did not. Later, mathematicians would use these discoveries about geometry to discover that there were other logics as well. Aristotle's system was, like Euclid's, just one of many possibilities. Even the concept of truth was not absolute. What is false in one logical system can be true in another. In Euclid's geometry of flat surfaces, parallel lines never meet, but on curved surfaces they can (see Figure 5.2).

These discoveries revealed the difference between mathematics and science. Mathematics was something bigger than science, requiring only self-consistency to be valid. It contained all possible patterns of logic. Some of those patterns were followed by parts of Nature; others were not. Mathematics was open-ended, uncompleteable, infinite; the physical Universe was smaller.

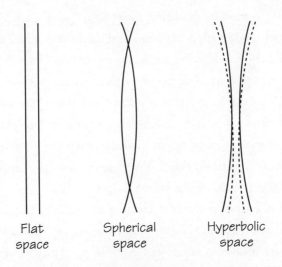

Flat space   Spherical space   Hyperbolic space

Figure 5.2 *Lines on flat and curved surfaces, where 'lines' are always defined by the shortest distance between two points. On a flat surface only parallel lines never meet; on the spherical surface all lines meet whilst on a hyperbolic space many lines never meet.*

MANY ZEROS

"The ultimate goal of mathematics is to eliminate all need for intelligent thought."

Ronald Graham, Donald Knuth & Owen Patashnik[9]

The discovery that there can exist logically self-consistent geometries which are different from Euclid's was a landmark.[10] It showed that mathematics was an infinite subject. There was no end to the number of different logical systems that could be invented. Some of those logical systems would have direct counterparts in the natural world, but others would not. Only a fraction of the possible patterns of mathematics are used in Nature.[11] From now on, some new choices would have to be made. What mathematical system is appropriate for the problem under study? If we wish to survey distances we need to use the right geometry. Euclid is no good for determining distances on the Earth's surface which are great enough for its curvature to be important.

The proliferation of mathematical systems (see Figure 5.3) led to the notion of what is now called 'mathematical modelling'. Particular pieces of mathematics help us describe aerodynamic motion but if we want to understand risk and chance we may have to use other mathematics. On the purer side of mathematics, it was recognised that there exist different mathematical structures, each defined by the objects (for example, numbers, angles or shapes) they contain and the rules for their manipulation (like addition or multiplication). These structures have different names according to the richness of the rules that are allowed.

One of the most important families of mathematical structures of this sort is that of a *group*. It is a precise prescription for a collection of objects that are related in some way. A group contains members, or 'elements', which can be combined by a transformation rule. This rule must possess three properties:

a. closure: if two elements are combined by the transformation rule, it must produce another element of the group.

b. identity: there must be an element (called the identity element)[12] which leaves unchanged any transformation it is combined with.

c. inversion: every transformation has an inverse transformation which undoes its effect on an element.

These three simple rules are based on properties that are possessed by many simple and interesting procedures. Let's consider a couple of examples. First, suppose that the group elements are all the positive and negative numbers (. . . $-3, -2, -1, 0, 1, 2, 3, \ldots$). The group transformation rule will be addition (+). We see that this defines a group because the closure condition is obeyed: the sum of any two numbers is always another number. The identity condition is obeyed. The identity element is zero, 0, and if we add it to any element it is left unchanged by +. The inversion property also holds: the inverse of the number N is $-N$ so that if we combine any number with its inverse we always get the identity, zero; for example $2 + (-2) = 2 - 2 = 0$.

Note that if we had taken our elements to be the same natural numbers but the transformation combining them to be multiplication rather than addition and the identity element to be 1, then the resulting structure is not a group. This is because the inversion property fails for all numbers other than $+1$ and -1. The quantity that we need to multiply, say, the number 3 by to give the identity, 1, is ⅓, which is not a whole number and so is not another element of the group. If we allow the elements to be fractions, then we do have a group with transformation defined by multiplication.

We notice that in these two examples the identity operation which leaves an element of the group unchanged is a null operation. In the first example of adding numbers it corresponds to the usual zero of arithmetic. Its status as the identity element of our group is guaranteed by the simple property that $N + 0 = N$ for any number N. In the second example the identity element is not the usual zero at all. The null operation for multiplication is provided by the number 1 (or, as a fraction ¼, which is the same thing). The usual zero is not a member of the second group.[13]

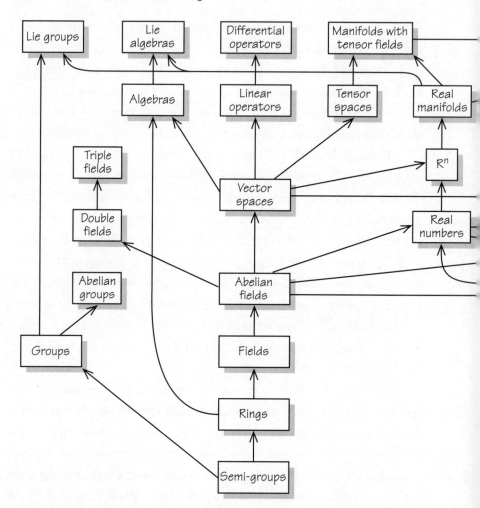

Figure 5.3 *The structure of modern mathematics, showing the development of different types of structure, from arithmetics, geometries and algebras. The simple natural numbers can be found at the heart of the network.*

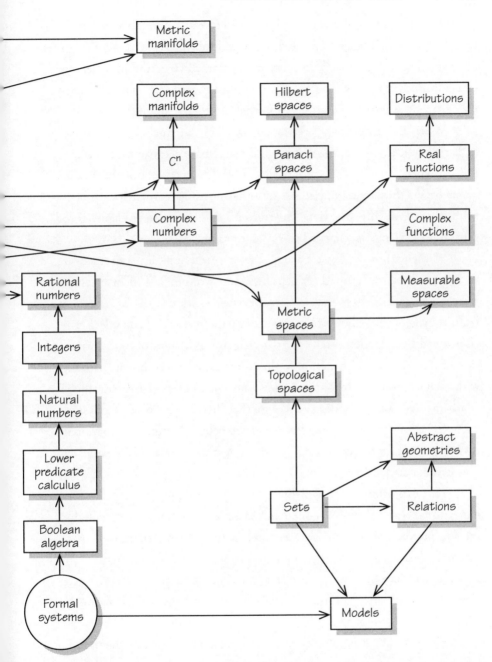

The elements in the second group structure are quite different from those in the first. The zero in the first group is quite distinct from that in the second group. Similarly, in every mathematical structure in which an element producing no change appears we must regard this 'zero' or 'identity' element as logically distinct from that in other structures.

When mathematicians were interested only in Euclidean geometry and arithmetic it was reasonable to regard mathematical existence and physical existence as being the same things. The discovery of non-Euclidean geometries, other logics and a host of other possible mathematical structures defined only by specifying the rules for combining their elements to generate new elements, changed this presumption.[14]

Mathematical existence parted company with physical existence. If the structure being invented on paper was free from logical inconsistency, then it was said to have mathematical existence. Its properties could be studied by exploring all the consequences of the prescribed rules. If a bad choice had been made initially for the elements and rules of transformation of a mathematical structure so that they turned out to be inconsistent with each other, then the structure was said not to exist mathematically.[15] Mathematical existence does not require that there be any part of physical reality that follows the same rules, but if we believe Nature to be rational then no part of physical reality could be described by a mathematically non-existent structure.

This explosion and fragmentation of mathematics (see Figure 5.3) has unusual consequences for the concept of zero. It creates a potentially infinite number of zeros. Each separate mathematical structure, fanned into mathematical existence by a judicious choice of a self-consistent set of axioms, may have its own zero element.[16] That zero element is defined solely by its null effect on the members of the mathematical structure in which it lives.[17]

The distinct nature of these zeros that inhabit different mathematical structures is nicely illustrated by an amusing paper written by Frank Harary and Ronald Read for a mathematics conference in 1973, entitled 'Is the null graph a pointless concept?'[18]

To a mathematician, a graph is a collection of points and lines joining some (or all) of the points. For example, a triangle made by joining up three points by straight lines is a simple 'graph' in this sense; so is the London Underground map. The null graph is the graph that possesses no points and no lines. It is shown in Figure 5.4.

There is a real difference between our old friend, the zero symbol, that the Indian mathematicians introduced long ago to fill the void in their arrays of numbers, and the zero or null operation that is needed to signify no change taking place in exotic mathematical structures. This zero operator is clearly something. It acts upon other mathematical objects; it follows rules; without it, the system is incomplete and less effective: it becomes a different structure.

This distinction between the traditional zero and other null mathematical entities is most spectacularly illustrated by the introduction of a definite notion of a collection, or a *set*, of things in mathematics. There is, as we shall see, a real and precise difference between the number zero and the concept of a set that possesses no members – the null, or empty, set.

Figure 5.4 *The null graph.*[19]

Indeed, the second idea, pointless as it sounds, turns out to be by far the most fruitful of the two. From it, all of the rest of mathematics can be created step by step.

CREATION OUT OF THE EMPTY SET

"A set is a set
(you bet; you bet!)
And nothing could not be a set,
you bet!
That is, my pet
Until you've met
My very special set."

Bruce Reznick[20]

One of the most powerful ideas in logic and mathematics has proved to be that of a *set*, introduced by the British logician George Boole. Boole was born in East Anglia in 1815 and is immortalised by the naming of Boolean logic/algebra/systems after him. He was responsible for the first revolution in human understanding of logic since the days of Aristotle. Boole's work appeared in a classic book, published in 1854, entitled *The Laws of Thought*.[21] It was then developed in important ways to deal with infinite sets by Georg Cantor between 1874 and 1897.

A set is a collection. Its members could be numbers, vegetables or individual's names. The set containing the three names Tom, Dick and Harry will be written as {Tom, Dick, Harry}. This set contains some simple subsets; for example, one containing only Tom and Dick {Tom, Dick}. In fact, it is easy to see that given any set we can always create a bigger set from it by forming the set which contains all the subsets of the first set.[22] The sets in this example have a finite number of members, but others, like that containing all the positive even numbers {2, 4, 6, 8, . . . and so on}, can have an infinite number of members generated by some rule.

Boole defined two simple ways of creating new sets from old. Given two sets A and B, the *union* of A and B, written A∪B, consists of all members of A together with all members of B; the intersection of A and B, written A∩B, is the set containing all the members common to both A and B. If A and B have no members in common they are said to be disjoint: their union is empty. These combinations are displayed in Figure 5.5.

One further idea is needed in order to use these notions. It is the concept of the *empty set* (or null set): the set that contains no members and is denoted by the symbol ∅, to distinguish it from our zero symbol, 0, of arithmetic. The distinction is clear if we think of the set of married bachelors. This set is empty, ∅, but the number of married bachelors in existence is zero, 0. We can also form a set of symbols whose only member is the zero symbol {0}.

We need the concept of the empty set to deal with the situation that arises when we encounter the intersection of two disjoint sets; for example, the set of all the positive even numbers and the set of all the positive odd numbers. They have no members in common and the set that is defined by their intersection is the empty set, the set with no members. This is the closest that mathematicians can get to nothingness. It seems rather different to the mystic or philosophical idea of nothingness which demands total non-existence. The empty set may have no members but it does seem to possess a degree of existence of the sort that sets have. It also possesses some similarities with the physical vacuum that we have already met. Just as the vacuum of nineteenth-century physics had the potential to be a part of everything, and has nothing inside it, so the empty set is the only set that is a subset of every other set.

All this sounds rather trivial but it turns out to have a remarkable pay-off. It allows us to define what we mean by the natural numbers in a simple and precise way by generating them all from nothing, the empty set. The trick is as follows.

Define the number zero, 0, *to be* the empty set, ∅, because it has no members. Now define the number 1 to be the set containing 0; that is, simply the set {0} which contains only one member. And, since 0 is defined to

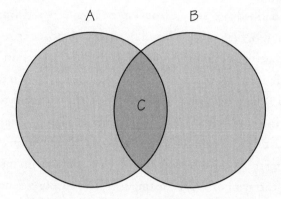

Figure 5.5 *Venn diagrams*[23] *illustrating the union and intersection (C) of two sets A and B.*

be the empty set, this means that the number I is the set that contains the empty set as a member $\{\varnothing\}$. It is important to see that this is by no means the same thing as the empty set. The empty set is a set with no members, whereas $\{\varnothing\}$ is a set containing one member.

Carrying on in this way we define the number 2 to be the set $\{0, I\}$, which is just the set $\{\varnothing, \{\varnothing\}\}$. Similarly the number 3 is defined to be the set $\{0, I, 2\}$ which reduces to $\{\varnothing, \{\varnothing\}, \{\varnothing, \{\varnothing\}\}\}$. In general, the number N is defined to be the set containing 0 and all the numbers smaller than N, so N = $\{0, I, 2, \ldots N-I\}$ is a set with N members. Every one of the numbers in this set can be replaced by their definition in terms of nested sets, like Russian dolls, involving only the concept of the empty set $\varnothing$. Despite the typographical nightmare this definition creates, it is beautifully simple in the way that it has enabled us to create all of the numbers from literally nothing, the set with no members.[24] This curious foundation for sets and numbers on the emptiness of the null set is nicely captured in a verse by Richard Cleveland:[25]

> "We can't be assured of a full set
> Or even a reasonably dull set.
>   It wouldn't be clear

That there's any set here,
Unless we assume there's a null set."

These strange sets within sets are mind-boggling at first. The incestuous way in which sets refer to themselves is not easy to get a feel for. But there is a more graphic way of visualising them[26] if we think about the part of our experience where the same self-reference constantly occurs — the process of thinking. Let's picture a set as a thought, floating in its thought balloon. Now just think about that thought. The empty set, $\varnothing$, is like an empty balloon but we can think *about* that empty thought balloon. This is like creating the set that contains the empty set $\{\varnothing\}$. This is what we called the number 1. Now go one further and think about yourself thinking about the empty set. This situation is $\{\varnothing, \{\varnothing\}\}$, which we call the number 2. By setting up this never-ending sequence of thoughts about thoughts, we produce an analogy for the definitions of the numbers from the empty set, as shown by the cartoons in Figure 5.6.

SURREAL NUMBERS

"In the beginning everything was void and J.H.W.H. Conway began to create numbers. Conway said, 'Let there be two rules which bring forth all numbers large and small. This shall be the first rule. Every number corresponds to two sets of previously created numbers, such that no member of the left set is greater than or equal to any member of the right set. And the second rule shall be this: One number is less than or equal to another number if and only if no member of the first number's left set is greater than or equal to the second number, and no member of the second number's right set is less than or equal to the first number.' And Conway examined these two rules he had made, and behold! they were very good."

Donald Knuth[27]

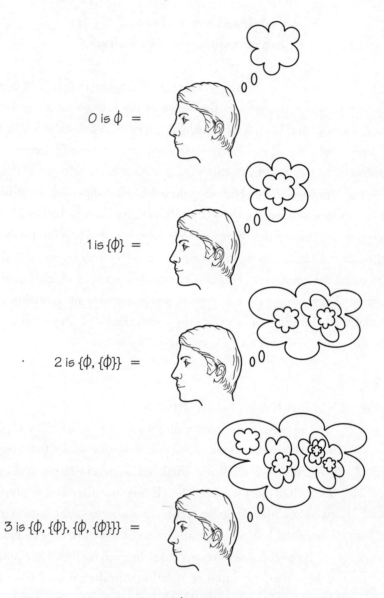

0 is ϕ =

1 is $\{\phi\}$ =

2 is $\{\phi, \{\phi\}\}$ =

3 is $\{\phi, \{\phi\}, \{\phi, \{\phi\}\}\}$ =

... and so on

Figure 5.6 *The mental analogy for the creation of the numbers from the empty set. A 'set' is represented by a thought and the empty set by an empty thought. Now think about that empty thought to generate the number 1, and so on.*

The fascination with using the empty set to create structure out of nothing at all didn't stop with the natural numbers. Quite recently, the ingenious English mathematician and maestro of logical games, John Conway, devised an imaginative new way of deriving not just the natural numbers, but the rational fractions, the unending decimals, and all other transfinite numbers as well, from an ingenious construction.[28] This population of children of nothing have been called the 'surreal' numbers by the computer scientist Donald Knuth,[29] who provided a novel exposition of the mathematical ideas involved by means of a fictional dialogue which traces his own exploration of Conway's ideas. Knuth has another serious purpose in mind in this story, besides explaining the mysteries of surreal numbers. He wants to make a point about how he believes mathematics should be taught and presented. Typical teaching lectures and textbooks are almost always a form of sanitised mathematics in which the intuitions and false starts that are the essence of the discovery process have been expunged.[30] The results are presented as a logical sequence of theorems, proofs and remarks. Knuth thinks that maths should be 'taken out of the classroom and into life', and he uses the surreal numbers as the prototype for this informal style of exposition. Here is something of the flavour of Conway's creation.

There are only two basic rules. First, every number (call it x) is made from two sets (a 'left set' L and a 'right set' R) of previously constructed numbers, so we write it down as[31]

$$x = \{L \,|\, R\}. \qquad (*)$$

These sets have the property that no member of the left set is greater than or equal to any member of the right set. Second, one number is less than or equal to another number if and only if no member of the first number's left set is greater than or equal to the second number, and no member of the second number's right set is less than or equal to the first number. The number zero can be created by choosing both the right and the left set to be the empty set, $\emptyset$, so

$$0 = \{\emptyset \,|\, \emptyset\}.$$

This definition follows the rules: first, no member of the empty set on the left is equal to or greater than any member of the right-hand empty set because the empty set has no members; second, 0 is less than or equal to 0. With a little thought, the rule can be extended to make the other natural numbers. We have $\emptyset$ and 0 to play with now and there are only two ways of combining them, which yield I and $-$I, respectively

$$I = \{\emptyset | 0\} \text{ and } -I = \{0 | \emptyset\}.$$

Carrying on in the same vein, we just put I and $-$I into the formula (*) and use it to generate all other natural numbers. Thus the positive number N allows us to generate N+I by combining it with the empty set through

$$\{N | \emptyset\} = N + I$$

and for the negative numbers we have

$$-N - I = \{\emptyset | -N\}.$$

Operations like addition and multiplication can also be defined self-consistently.[32] The empty set behaves in a simple way. The empty set plus anything is still just the empty set and the empty set multiplied by anything else is still the empty set.

Again, this is all very pretty but what does it enable us to do that we couldn't do with the old scheme that we discussed above? The pay-off comes when Conway extends his scheme to include more exotic numbers in the L and R slots. For example, suppose one takes the set L to be an infinity of natural numbers (called a *countable* infinity) 0, I, 2, 3, . . . and so on, for ever. Then we can define infinity to be[33]

$$\text{inf} = \{0, I, 2, 3, \ldots | \emptyset\}$$

Now put inf in the right-hand slot and we have a peculiar definition for infinity minus I, an infinite number less than infinity!,

$$\text{inf} - I = \{0, I, 2, 3, \ldots | \text{inf}\}$$

and also

$$I/\text{inf} = \{0 | \tfrac{1}{2}, \tfrac{1}{4}, \tfrac{1}{8}, \tfrac{1}{16}, \ldots\}$$

and even, the square root of infinity:

$$\sqrt{\text{inf}} = \{0, I, 2, 3, \ldots | \text{inf}, \tfrac{\text{inf}}{2}, \tfrac{\text{inf}}{4}, \tfrac{\text{inf}}{8} \ldots\}.$$

None of these peculiar quantities had been defined by mathematicians previously. Starting from the empty set and two simple rules, Conway manages to construct all the different orders of infinity found by Cantor, as well as an unlimited number of strange beasts like $\sqrt{\text{inf}}$ that had not been defined before. Every real decimal number that we know finds itself surrounded by a cloud of new 'surreal' numbers that lie closer to it than does any other real number. Thus the whole of known mathematics, from zero to infinity, along with unsuspected new numbers hiding in between the known numbers, can be created from that seeming nonentity, the empty set, ∅. Who said that only nothing can come of nothing?

GOD AND THE EMPTY SET

"You know the formula: m over nought equals infinity, m being any positive number? Well, why not reduce the equation to a simpler form by multiplying both sides by nought? In which case, you have m equals infinity times nought. That is to say that a positive number is the product of zero and infinity. Doesn't that demonstrate the creation of the universe by an infinite power out of nothing?"

Aldous Huxley[34]

Our discussion of the unexpected richness of the empty set leads us to take a look at its relationship to the infamous ontological argument for the existence of God.[35] This argument was first propounded by Anselm, who was Archbishop of Canterbury, in 1078. Anselm conceives[36] of God as something than which nothing greater or more perfect can be conceived. Since this idea arises in our minds it certainly has an intellectual existence. But does it have an existence outside of our minds? Anselm argued that it must, for otherwise we fall into a contradiction. For we could imagine something greater than that which nothing greater can be conceived; that is the mental conception we have together, plus the added attribute of real existence.

This argument has vexed philosophers and theologians down the centuries and it is universally rejected by modern philosophers, with the exception of Charles Hartshorne.[37] The doubters take their lead from

Kant, who pointed out that the argument assumes that 'existence' is a property of things whereas it is really a precondition for something to have properties. For example, while we can say that 'some white tigers exist', it is conceptually meaningless to say that 'some white tigers exist, and some do not'. This suggests that while whiteness can be a property of tigers, existence cannot. Existence does not allow us to distinguish (potentially) between different tigers in the way that colour does. Despite its grammatical correctness, it is not logically correct to assert that because something is a logical possibility, it must necessarily exist in actuality.

We see that there is an amusing counterpart to these attempts to prove that God, defined as the greatest and most perfect being, necessarily exists because otherwise He would not be as perfect as He could be. For suppose that the empty set, conceived as that set than which no emptier set can be conceived, did not exist. Then we could form a set that contained all these non-existent sets. This set would be empty and so it is necessarily the empty set! One can see that with a suitable definition of the Devil as something than which nothing less perfect can be conceived, we could use Anselm's logic to deduce the non-existence of the Devil since a non-existent Devil has a lower status than one which possesses the attribute of existence.

LONG DIVISION

"Now: heaven knows, anything goes."

Cole Porter

The mathematical developments we have charted in this chapter show how a great divide came between the old nexus of zero, nothingness and the void. Once, these ideas were part of a single intuition. The rigorous mathematical games that could be played with the Indian zero symbol had given credibility to the philosophical search for a meaningful concept of how nothing could be something. But in the end mathematics was too great an empire to remain intimately linked to physical reality. At first, mathematicians took their ideas of counting and geometry largely from the world around them. They believed there to be a single geometry and a single logic. In the nineteenth century they began to see further. These simple systems of mathematics they had abstracted from the natural world provided models from which new abstract structures, defined solely by the rules for combining their symbols, could be created. Mathematics was potentially infinite. The subset of mathematics which described parts of the physical universe was smaller, perhaps even finite. Each mathematical structure was logically independent of the others. Many contained 'zeros' or 'identity' elements. Yet, even though they might share the name of zero, they were quite distinct, having an existence only within the mathematical structure in which they were defined and logically underwritten by the rules they were assumed to obey. Their power lay in their generality, their generality in their lack of specificity. Bertrand Russell, writing in 1901, captured its new spirit better than anyone:

> "Pure mathematics consists entirely of such asseverations as that, if such and such a proposition is true of *anything*, then such and such another proposition is true of that thing. It is essential not to discuss whether the first proposition is really true, and not to mention what the anything is of which it is supposed to be true . . . If our hypothesis is about *anything* and not about some one or more particular things, then our deductions constitute mathematics. Thus mathematics may be defined as the subject in which we never know what we are talking about, nor whether what we are saying is true."[38]

Pure mathematics became the first of the ancient subjects to free itself of metaphysical shackles. Pure mathematics became free mathematics. It could invent ideas without recourse to correspondence with anything in the worlds of science, philosophy or theology. Ironically, this renaissance emerged most forcefully not with the plurality of zeros that it spawned, but with the plethora of infinities that Georg Cantor unleashed upon the unsuspecting community of mathematicians. The ancient prejudice that there could be potential infinities, but never actual infinities, was ignored. Cantor introduced infinities without end in the face of howls of protest by conservative elements in the world of mathematics. Cantor was eventually driven into the deep depression that overshadowed the end of his life, yet he vigorously maintained the freedom of mathematicians to invent what they will:

> "Because of this extraordinary position which distinguishes mathematics from all other sciences, and which produces an explanation for the relatively free and easy way of pursuing it, it especially deserves the name of *free mathematics*, a designation which I, if I had the choice, would prefer to the customary 'pure' mathematics."[39]

These free-spirited developments in mathematics marked the beginning of the end for metaphysical influences on the direction of the mathematical imagination. Nothingness was unshackled from zero, leaving the vagueness of the void and the vacuum behind. But there were more surprises to come. The exotic mathematical structures emerging from the world of pure mathematics may have been conceived free from application to Nature, but something wonderful and mysterious was about to happen. Some of those same flights of mathematical fancy, picked out for their symmetry, their neatness, or merely to satisfy some rationalist urge to generalise, were about to make an unscheduled appearance on the stage of science. The vacuum was about to discover what the application of the new mathematics had in store for time and space and all that's gone before.

chapter six

Empty Universes

"You cannot have first space and then things to put into it, any more than you can have first a grin and then a Cheshire cat to fit on to it."

Alfred North Whitehead

DEALING WITH ENTIRE UNIVERSES ON PAPER

"I always think love is a little like cosmology. There's a Big Bang, a lot of heat, followed by a gradual drifting apart, and a cooling off which means that a lover is pretty much the same as any cosmologist."

Philip Kerr[1]

The most spectacular intellectual achievement of the twentieth century is Einstein's theory of gravity. It is known as the 'general theory of relativity' and supersedes Newton's three-hundred-year-old theory. It is a generalisation of Newton's theory because it can be used to describe systems in which objects move at a speed approaching that of light and in gravitational fields which are extremely strong.[2] Yet, when applied to environments where speeds are low and gravity is very weak, it looks like Newton's theory. In our solar system the distinctive differences between Newton and Einstein are equal to just one part in one hundred thousand, but these are easily detectable by astronomical instruments. Far from Earth, in high-density astronomical environments, the differences between the predictions of Newton and Einstein are vastly larger, and so far our observations have confirmed Einstein's predictions to an accuracy that exceeds the confirmation of any other scientific theory. Remarkably, the picture that Einstein has given us of the way in which gravity behaves, locally and cosmically, is the surest guide we have to the structure of the Universe and the events that occur within it.

From this short prologue one could be forgiven for thinking that Einstein's gravity theory is just a small extension of Newton's, a little tweaking of his claim that the force between two masses falls off in proportion to the square of the distance separating their centres. Nothing could be further from the truth. Although in some situations the differences between the predictions of Einstein and Newton are very small, Einstein's conceptions of space and time are radically different. For Newton, space and time were absolutely fixed quantities, unaffected by the presence of the bodies contained within them. Space and time provided the arena in which motion took place; Newton's laws gave the marching orders.

When gravity attracted different masses, it was supposed to act instantaneously through the space between them regardless of their separation. No mechanism was proposed by which this notorious 'action at a distance' could occur. Newton was as aware of this lacuna as anybody, but pushed ahead regardless with his simple and successful law of gravity because it worked so well, giving accurate predictions of the tides, the shape of the Earth, together with an explanation of many observed lunar, astronomical and terrestrial motions. Indeed, one could go along sweeping this problem under the rug, secure in the knowledge that it wasn't creating any crises for human thought elsewhere, right up until the discovery of the special theory of relativity. Relativity predicted that it should be impossible to send information faster than the speed of light in a vacuum.[3]

In 1915, Einstein solved the problem of how gravity acts in a novel way. He proposed that the structure of space and time is not fixed and unchanging like a flat table top; rather, it is shaped and distorted by the presence of mass and energy[4] distributed within it. It behaves like a rubber sheet that forms undulations when objects are placed upon it. When mass and energy are absent, the space is flat. As masses are added, the space curves. If the masses are large, the distortion of the flat surface of space is large near to the mass but decreases as one moves far away. This simple analogy is quite suggestive. It implies that if we were to wiggle a mass up and down at a point on the rubber sheet so as to produce ripples, as on the surface of a pool of water, then the ripples would travel outwards like waves of gravity. Also, if one were to rotate a mass at one point on the rubber sheet, it would twist the sheet slightly, further away, dragging other masses around in the same direction. Both these effects occur in Einstein's theory and have been observed.[5] Einstein discovered two important sets of mathematical equations. The first, called the 'field equations', enable you to calculate what the geometry of space and time[6] is for any particular distribution of matter and energy within it. The other, called the 'equations of motion', tell us how objects and light rays move on the curved space. And what they tell is beautifully simple. Things move so that they take the quickest route over the undulating surface prescribed by the field equations. It is like following the path taken by a stream that meanders down from the mountain top to the river plain below.

This picture of matter curving space and curvaceous space dictating how matter and light will move has several striking features. It brings the non-Euclidean geometries that we talked about in the last chapter out from the library of pure mathematics into the arena of science. The vast collection of geometries describing spaces that are not simply the flat space of Euclid are the ones that Einstein used to capture the possible structures of space distorted by the presence of mass and energy. Einstein also did away with the idea of a gravitational force (although it is so ingrained in our intuitions that astronomers still use it as a handy way of describing the appearances of things), and with it the problematic notion of its instantaneous action at a distance. You see, in Einstein's vision, the motions of bodies on the curved space are dictated by the local topography that they encounter. They simply take the quickest path that they can. When an asteroid passes near the Sun it experiences a region where the curvature of space is significantly distorted by the Sun's presence and will move towards the Sun in order to stay on a track that will minimise its transit time (see Figure 6.1). To an observer just comparing their relative positions it looks as if the planet is attracted to the Sun by a force. But Einstein makes no

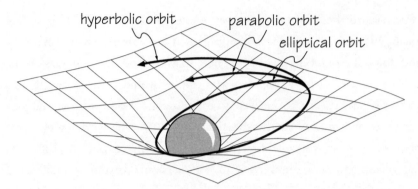

Figure 6.1 *Bodies that move take the quickest route between two points on a curved surface.*

mention of any forces: everything moves as if acted upon by no forces and so moves along a path that is the analogue of a straight line in flat Euclidean space. Moving objects take their marching orders from the local curvature of space, not from any mysterious long-range force of gravity acting instantaneously without a mechanism.

Einstein's theory had a number of spectacular successes soon after it was first proposed. It explained the discrepancy between the observed motion of the planet Mercury and that predicted by Newton's theory, and successfully predicted the amount by which distant starlight would be deviated by the Sun's gravity en route to our telescopes. Yet its most dramatic contribution to our understanding of the world was the ability it gave us to discuss the structure and evolution of entire universes, even our own.

Every solution of Einstein's field equations describes an entire universe – what astronomers sometimes call a 'spacetime'. At each moment of time a solution tells us what the shape of space looks like. If we stack up those curved slices then they produce an unfolding picture of how the shape of space evolves in response to the motion and interaction of the mass and energy it contains. This stack is the spacetime.[7] The field equations tell us the particular map of space and the pattern of time change created by a given distribution of mass and energy. Thus a 'solution' of the equations gives us a matching pair: the geometry that is created by a particular distribution of mass and energy, or conversely, the curved geometry needed to accommodate a specified pattern of mass and energy. Needless to say, Einstein's field equations are extremely difficult to solve and the solutions that we know always describe a distribution of matter and a geometry that has certain special and simplifying properties. For example, the density of matter might be the same everywhere (we say it is homogeneous in space), or the same in every direction (we say it is isotropic), or assumed to be unchanging in time (static) . If we don't make one of these special assumptions we have to be content with approximate solutions to the equations which are valid when the distribution is 'almost' homogeneous, 'almost' isotropic, almost 'static' or changes in a very simple way (rotating at a steady speed, for example). Even these simpler situations are mathematically very complicated and make Einstein's theory extremely difficult to use in all the ways one would like. Often, supercomputing capability is required to carry out studies of how very realistic configurations, like pairs of stars, will behave. This complexity is, however, not a defect of Einstein's theory in any sense. It is a reflection of the complexity of gravitation. Gravity acts on all forms of mass and energy, but energy comes in a host of very different forms that behave in peculiar ways that were not known in Newton's day. Worst of all, gravity gravitates. Those waves of gravity that spread out, rippling the curvature of space, carry energy too and that energy acts as a source for its own gravity field. Gravity interacts with itself in a way that light does not.[8]

VACUUM UNIVERSES

"... and he shall stretch out upon it the line of confusion, and the stones of emptiness. They shall call the nobles thereof to the kingdom, but none shall be there."

Isaiah[9]

The fact that the solutions of Einstein's theory describe whole universes is striking. Some of the first solutions that were found to his field equations provided excellent descriptions of the astronomical universe around us that telescopes would soon confirm. They also highlighted a new concept of the vacuum.

We have seen that Einstein's equations provide the recipe for calculating the curved geometry of space that is created by a given distribution of mass and energy in the Universe. From this description one might have expected that if there were no matter or energy present – that is, if space was a perfectly empty vacuum in the traditional sense – then space would be flat and undistorted. Unfortunately, things are not so simple. A geometry that is completely flat and undistorted is indeed a solution to the equations when there is no mass and energy present, as one would expect. But there are many other solutions that describe universes containing neither mass nor energy but which have *curved* spatial geometries.

These solutions of Einstein's equations describe what are called 'vacuum' or 'empty' universes. They describe universes with three dimensions of space and one of time, but they can be imagined more easily if we forget about one of the dimensions of space and think of worlds with just two dimensions of space at any moment of time, like a table top, but not necessarily flat, so rather like a trampoline. As time flows the topography of the surface of space can change, becoming flatter or more curved and contorted in some places. At each moment of time we have a different 'slice' of curved space.[10] If we stack them all up in a pile then we create the whole spacetime, like making a lump of cheese out of many thin slices (see Figure 6.2). If one picked any old collection of slices and stacked them up, they would not fit together in a smooth and natural way that would correspond to a smooth flow of events linked by a chain of causes and effects. That's where Einstein's equations come in. They guarantee that this stacking will make sense if the ingredients solve the equations.[11]

This is all very well, but having got a picture of how Einstein's theory works by imagining that the presence of mass and energy creates curvatures in the geometry of space and changes in the rate of flow of time, shouldn't empty universes all be flat? If they contain no stars, planets and atoms of matter, how can space be curved? What is there to do the curving?

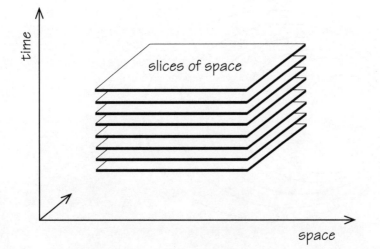

Figure 6.2 *Spacetime is composed of a stack of slices of space, each one labelled by a moment of time. Only two of the dimensions of space are shown.*

Einstein's theory of gravity is much larger than Newton's. It does away with the idea that the effects of gravity are instantaneously communicated from one side of the Universe to the other and incorporates the restriction that information cannot be sent at speeds faster than that of light. This allows gravity to spread its influence by means of waves travelling at the speed of light. These gravitational waves were predicted to exist by Einstein and there is little doubt that they do exist. Although they are too weak to detect directly on Earth today, their indirect effects have been observed in a binary star system containing a pulsar. The pulsar is like a lighthouse beam spinning at high speed. Every time it comes around to face us we see a flash. Its rotation can be very accurately monitored by timing observations of its periodic pulses. Twenty years of observations have shown that the pulsing of the binary pulsar is slowing at exactly the rate predicted if the system is losing energy by radiating gravitational waves at the rate predicted by Einstein's theory (see Figure 6.3).

In the next few years, ambitious new experiments will attempt to detect these waves directly. They are like tidal forces in their effects. When a gravitational wave passes through the page that you are reading it will slightly stretch the book sideways and squeeze it longways without

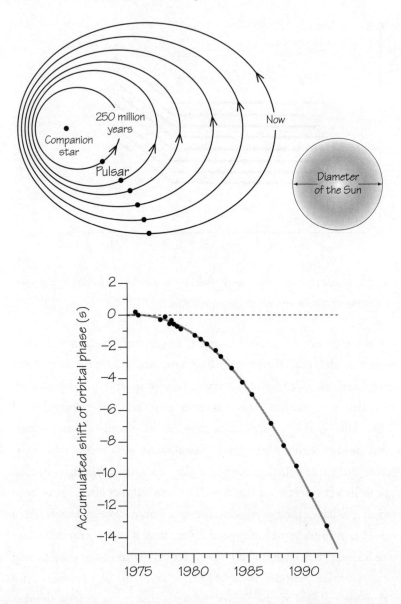

Figure 6.3 *The Binary Pulsar PSR 1913+16, one of about 50 known systems of this type. It contains two neutron stars orbiting around one another. One of the neutron stars is a pulsar and emits pulses of radio waves which can be measured to high precision. These observations show that the orbital period of the pulsar is changing by 2.7 parts in a billion per year. This is the change predicted by the general theory of relativity due to the loss of energy by the radiation of gravitational waves from the neutron stars.*[12]

changing its volume. The effect is tiny but with elaborate apparatus, similar to the interferometer used by Michelson to test the existence of the ether, we may be able to detect gravitational waves from violent events far away in our own galaxy and beyond. The prime candidates for detection are waves from very dense stars or black holes that are in the final throes of circling each other in orbits that are getting closer and closer to each other. In the end they will spiral together and merge in a cataclysmic event that produces huge amounts of outgoing light and gravitational waves. In the far future the binary pulsar will collapse into this state and provide a spectacular explosion of gravitational waves.

If we imagine a space that has had its geometry distorted by the presence of a large mass then we can see how gravitational waves can alter the picture. Suppose that the mass starts changing shape in a way that makes it non-spherical. The changes create ripples in the geometry which spread out through the geometry, moving away from the mass. The further away one is from the source of this disturbance the weaker will be the effect of the ripples when they reach you. Although we talk about these waves as if they are a form of energy, like sound waves, that have been introduced into the Universe, they are really rather different in character. They are an aspect of the geometry of space and time. If we take away the changing mass that is generating the ripples in the geometry of space we can still have such ripples present. The whole Universe can be expanding in a non-spherical way, slightly faster in one direction than in another, and very long-wavelength gravitational waves will be present to support the overall tension in the expanding 'rubber sheet' that is the universe of space at any moment.

The fact that Einstein's theory of gravity allows one to find precise descriptions of universes which expand, like our own, but which contain no matter does not mean that such universes are realistic. Einstein's theory is remarkable in that it describes an infinite collection of possible universes of all shapes and sizes according to the distribution and nature of the matter you care to put into them. One of the simplest solutions of Einstein's equations, which does contain matter, and expands at the same rate in every direction and at every place, gives an extremely accurate description of the behaviour of our observed Universe. The biggest problem facing cosmologists is to explain why this solution is selected to pass from being a mere theoretical possibility to real physical existence. Why this simple universe and not some other solution of Einstein's equations?

We expect that there is more to the Universe than we can learn from Einstein's equations alone. Linking Einstein's theory up to our understanding of the most elementary particles of matter may place severe restrictions on which curved spaces are physically possible. Or it may be that strange forms of matter existed in the early stages of the Universe's history which ensure that all, or almost all, of the complicated universes which Einstein's equations permit end up looking more and more like the simple isotropically expanding state that we see today if you wait for billions of years.

ERNST MACH – A MAN OF PRINCIPLE

"It is easy to be certain. One has only to be sufficiently vague."

Charles Sanders Peirce[13]

Einstein's own views about empty universes played an important role in his conception and creation of the general theory of relativity. Over a long period of time, his thinking about the defects in Newton's theory and how to repair them had been much influenced by the physicist and philosopher of science, Ernst Mach (1838–1916). Mach had wide interests and made important contributions to the study of sound. Aerodynamicists invariably label high velocities by their 'Mach number', that is the value of the speed in units of the speed of sound (about 750 miles per hour). Yet in some respects Mach was something of a Luddite and opposed the concept of atoms and molecules as basic components of matter on philosophical grounds even after there was direct experimental evidence for their existence. Nonetheless, Einstein had been greatly impressed by Mach's famous text on mechanics[14] and it played an important role in guiding him to formulate both the special and general theories of relativity in the ways that he did. As a consequence, Einstein was much influenced by another of Mach's convictions about the origin of the inertia of local objects, a view that has since become known as 'Mach's Principle'. Mach believed that the inertia and mass of the objects we see around us should be a consequence of the collective effect of the gravitational field of all the mass in the Universe. When Einstein conceived of his general theory of relativity, with the presence of mass and energy creating curvature, he hoped and believed that Mach's idea was automatically built into it. Alas, it was not. Mach's Principle boiled down to requiring that there were no vacuum solutions of Einstein's theory: no universes where the geometry of space and time was curved by gravitational waves alone, rather than by the presence of mass and energy. Gravitational waves were allowed to exist, but they had to arise from the movement of irregular distributions of matter. There could not exist wavelike ripples in the geometry of space that were built into the Universe when it came into being or which were associated solely with disparities in its expansion rate from one direction to another.

The most dramatic type of motion of the Universe, not associated with matter, that Mach and Einstein needed to veto was an overall cosmic rotation. For a long time Einstein believed that his theory ensured this, but he got a surprise. In 1952, the logician Kurt Gödel, his colleague at the Institute for Advanced Study in Princeton, discovered a completely unexpected solution of Einstein's equations that described a rotating universe. More dramatic still, this possible universe permitted time travel to take place! Subsequent investigation showed that this solution of Einstein's equations was peculiar and could not describe our Universe. However, the genie was out of the bottle. Maybe there were other solutions which possessed the same properties but which were much more realistic? Or maybe Mach was right and we just haven't found the right way to formulate his 'Principle' when we look for solutions of Einstein's equations. For no overall rotation of the Universe has ever been detected. Some years ago, some of us[15] used astronomical observations of the isotropy of the intensity of radiation in the Universe to show that if it is rotating then it must be rotating at a rate that is between one million and ten million million times slower than the rate at which the Universe is expanding.[16]

Mach's Principle reflects older ideas about the undesirability of a vacuum. It is largely ignored in modern cosmology, not least because it is rather difficult to get everyone to agree on a precise statement of the Principle. Many scientists have tried to modernise it to see if it can be used as a way of selecting out some of the solutions of Einstein's equations as physically realistic but none of the proposals has caught on. Even if they did, it is not clear what Mach's Principle would tell us that we could not learn in other ways. It is all very well to say that gravitational fields must all arise from sources of matter, with no free gravitational waves left over from the Big Bang, but *why* should such a state of affairs exist? If our Universe was dominated by the presence of very strong sourceless gravitational waves then its expansion would behave very differently. It would expand at quite different rates in different directions and it might rotate as fast as it expands. Our observations show us that neither of these scenarios exists in the Universe today. The expansion of the Universe is the same in every direction to an accuracy of one part in one hundred thousand.

Mach's Principle faded from the stage because it could not supply an answer to the question 'Why is the Universe like it is today?' Later, we shall see that other ideas have been able to come up with more compelling reasons for the lack of measurable effects by sourceless gravitational waves in the universes today. They do not stipulate that those waves cannot exist, as Mach would have decreed, but show that they are inevitably very weak when the universe is old and have negligible effects upon the overall expansion of the Universe.

LAMBDA – A NEW COSMIC FORCE

"If an elderly but distinguished scientist says that something is possible he is almost certainly right, but if he says that it is impossible he is very probably wrong."

Arthur C. Clarke

When Albert Einstein first began to explore the cosmological consequences of his new theory of gravity, in 1915, our knowledge of the scale
and diversity of the astronomical universe was vastly smaller than it is
today. There was no reason to believe that there existed galaxies other than
our own Milky Way. Astronomers were interested in stars, planets, comets
and asteroids. Einstein wanted to use his equations to describe our whole
Universe but they were too complicated for him to solve without some
simplifying assumptions. Here he was very fortunate. He assumed something about the Universe that certainly makes life easy for the mathematician but which might well not have been an appropriate assumption to
make about the real Universe. The observational evidence simply did not
exist. Einstein's simplifying assumption was that the Universe is the same in
every place and in every direction at any moment of time. We say that it is
homogeneous and isotropic. Of course, it is not exactly so. But the assumption is that it is so close to being so that the deviations from perfect uniformity are too small to make any significant difference to the mathematical
description of the whole Universe.[17]

   As Einstein continued he found that his equations were telling him
something very peculiar and unexpected: the Universe had to be constantly
changing. It was impossible to find a solution for a universe which contained a uniform distribution of matter, representing the distant stars,
which remained on average the same for long periods of time. The stars
would attract one another by the force of their gravity. In order to avoid a
contraction and pile-up of matter in a cosmic implosion, there would need
to be an outward motion of expansion to overcome it – an 'expanding' universe.

   Einstein didn't like either of these alternatives. They were both contrary to the contemporary conception of the Universe as a vast unchanging
stage on which the motions of the celestial bodies were played out. Stars
and planets may come and go, but the Universe should go on for ever.
Faced with this dilemma of a contracting or an expanding universe, he
returned to his equations and searched for an escape clause. Remarkably, he
found one.

   To see how this happened we must first see something of what led

Einstein to his original equations. His equations relating the geometry of curved space to the material content of space have a particular form:

$$\{\text{geometry}\} = \{\text{distribution of mass and energy}\}.$$

All sorts of formulae describing the shapes of surfaces are possible in principle on the left-hand side of this equation. But if they are going to be equated to realistic distributions of matter and radiation, with properties like density, velocity and pressure, then they must reflect the fact that quantities like energy and momentum have to be conserved in Nature. They can be reshuffled and redistributed in all sorts of ways when interactions occur between different objects, but when all the changes are complete and all the energies and momenta are finally added up they must give the same sums that they did at the start. This requirement, that energy and momentum be conserved in Nature, was enough to guide Einstein to the simplest geometrical ingredients on the left-hand side of his equations.

Everything seemed to fit together beautifully. If he looked at the situation where gravity was very weak and speeds were far less than that of light, so that the deviations in the geometry of space from perfect Euclidean flatness were tiny, then these complex equations miraculously turned into the self-same law of gravity that Newton had discovered more than 230 years earlier. This law was called the 'inverse-square law' because it dictated that the gravitational force between two masses falls inversely as the square of the distance between their centres.

Unfortunately, it was this elegant picture that stubbornly refused to allow the Universe to be unchanging. Faced with an expanding universe, Einstein saw a way out. His desire to make his theory turn into Newton's when gravity became very weak, and space was nearly flat, had led him to ignore a strange possibility. The parts of his equations storing the information about the geometry allowed another simple piece to be added to them without altering the requirement that they allow energy and momentum to be conserved in Nature. When one looked at what this new addition would do to Newton's description of weak gravitational fields, the result looked very odd. It said that Newton's inverse-square law was only half the story; there was really another piece to be added to it: a force between all masses that *increased* in proportion to the distance of their separation. As one looked out to astronomical distances this extra force of gravity should overwhelm the effects of Newton's decreasing inverse-square law.

Einstein introduced the Greek symbol lambda, Λ, to denote the strength of this force in his equations, so that schematically they became:

{geometry} + {Λ force} = {distribution of mass and energy}.

Nothing in his theory could tell him how large a number lambda was, or even whether lambda was positive or negative. Indeed, an important reason to keep it in his equations was that, equally, there was no reason why its value should be zero either. Lambda was a new constant of Nature, like Newton's gravitation constant, G, which determined the strength of the attractive, inverse-square part of the gravitational force. Einstein called lambda the 'cosmological constant'.

Einstein saw that if lambda was positive then its repulsive contribution to the overall force of gravity would be opposite to the attractive character of Newton's force. It would cause distant masses to repel one another. He realised that if its value was chosen appropriately it could exactly counterbalance the gravitational attraction of the inverse-square law and so allow a universe of stars to be static, neither expanding nor contracting. The fact that we did not see any evidence on Earth for this lambda force was easily explained. The value of lambda required to keep the Universe static was very small, so small that its consequences on Earth would be far too small to have any perceptible effect on our measurements of gravity. This situation arose because the force increased with distance. It could be large over astronomical dimensions where it controlled the overall stability of the Universe, yet be very small over the small distances encountered on the surface of the Earth or in the solar system.

What happened next was something of an embarrassment for Einstein. He believed that his static universe was the only type of solution that his new equations permitted for the Universe. However, he was not the only person studying his equations.

Alexander Friedmann was a young meteorologist and applied mathematician working in St Petersburg. He followed new developments in mathematical physics closely and was one of the very first scientists to understand the mathematics behind Einstein's new theory of gravity. This was a remarkable achievement. Einstein's theory used parts of mathematics that were highly abstract and which had never been used in physics before. Astronomers were, for the most part, practically inclined physicists rather than mathematical specialists, and ill equipped to understand Einstein's theory at a level that enabled them to check his calculations and go on and do new ones. Friedmann was different. He assimilated the mathematics required very quickly and was soon finding new solutions of Einstein's equations which Einstein himself had missed.[18] He found the expanding and contracting solutions that Einstein had tried to suppress by introducing the lambda term. The three varieties of expanding universe are shown in Figure 6.4. But he also found something more interesting. Even with Einstein's lambda force added to the equations, the Universe would not remain static. The solution that Einstein found in which the attractive force of gravity exactly balanced the new repulsive lambda force did exist. But it would not persist. It was unstable. Like a needle balanced on its point, if nudged in any direction, it would fall. If Einstein's static universe possessed the slightest irregularity in its density, no matter how small, it would begin to expand or contract. Friedmann confirmed this by showing that even when the lambda force was present there were solutions to Einstein's equations which described expanding universes. Following these calculations to their logical conclusion, Friedmann made the greatest scientific prediction of the twentieth century: that the whole Universe should be expanding.[19]

Friedmann wrote to Einstein to tell him that there were other solutions to his equations but Einstein didn't pay close attention, believing Friedmann's calculations to be mistaken. Soon afterwards one of Friedmann's more senior colleagues went to Berlin on a lecture tour with the added purpose of discussing Friedmann's calculations with Einstein. Einstein was rapidly persuaded that it was he, rather than Friedmann, who was mistaken; he had completely overlooked the new solutions to his equations. Einstein wrote to announce that Friedmann was correct and the static universe was dead. Years later, Einstein would describe his invention of the cosmological constant to sustain his belief in a static universe as 'the biggest blunder of my life'.

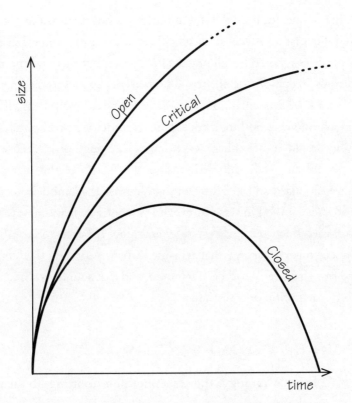

Figure 6.4 *The three universes discovered by Friedmann. The open and critical cases increase in spatial extent for ever; the closed case eventually collapses back to a state of maximum compression. The critical trajectory is the dividing line between infinite and finite future histories.*

In 1929, astronomers finally established that the Universe is indeed expanding, just as Friedmann had predicted, and Friedmann's solutions of Einstein's equations, both with and without the lambda force, still provide the best working descriptions of that expansion today. Friedmann never lived to see how far-reaching his ideas had been. Tragically, he died when only thirty-five years old after failing to recover from the effects of a high-altitude balloon flight to gather meteorological data.[20]

Despite the debacle of the static universe, the lambda force lived on. Einstein's logic that led to its inclusion in his equations was inescapable, even if the desideratum of a static universe was not. Lambda might be so small in value that its effects are negligible even over astronomical distances and its presence ignored for all practical purposes, but there was no reason just to leave it out of the theory. Observations soon showed that if it existed it must be very small. But why should it be so small? Einstein's theory told astronomers nothing about its magnitude or its real physical origin. What could it be? These were important questions because their answers would surely tell us something about the nature of the vacuum. For even if we expunged all the matter in the Universe the lambda force could still exist, causing the Universe to expand or contract. It was always there, acting on everything but unaffected by anything. It began to look like an omnipresent form of energy that remained when everything that could be removed from a universe had been removed, and that sounds very much like somebody's definition of a vacuum.

DEEP CONNECTIONS

"I love cosmology: there's something uplifting about viewing the entire universe as a single object with a certain shape. What entity, short of God, could be nobler or worthier of man's attention than the cosmos itself? Forget about interest rates, forget about war and murder, let's talk about *space*."

Rudy Rucker[21]

The first person to suggest that the cosmological constant might be linked to the rest of physics was the Belgian astronomer and Catholic priest, Georges Lemaître. Lemaître was one of the first scientists to take the idea of the expanding Universe seriously as a problem of physics. If the Universe was expanding, then he realised that it must have been hotter and denser in the past: matter would be transformed into heat radiation if cosmic events were traced far enough into the past.

Lemaître rather liked Einstein's lambda force and found several new solutions of Einstein's equations in which it featured. He was persuaded that it needed to be present in Einstein's theory but, unlike Einstein, who tried to forget about it, and some other astronomers, who assumed that even if it existed it was negligible, he wanted to reinterpret it. Lemaître realised[22] that although Einstein had added the lambda force to the geometrical side of his equations, it was possible to shift it across to the matter and energy side of the equation

{geometry} = {distribution of mass and energy} − {Λ energy},

and reinterpret it as a contribution to the material content of the Universe,

{geometry} = {distribution of mass and energy − Λ mass and

Λ energy}.

If you do this then you have to accept that the Universe always contained a strange fluid whose pressure is equal to *minus* its energy density. A negative pressure is just a tension which is not unusual, but the lambda tension is as negative as it could possibly be and this means that it exerts a gravitational effect that is *repulsive.*[23]

Lemaître's insight was very important because he saw that by interpreting the cosmological constant in this way it might be possible to understand how it originated by studying the behaviour of matter at very high energies. If those investigations could identify a form of matter which existed with this unusual relation between its pressure and its energy density then it would be possible to link our understanding of gravity and the geometry of the Universe to other areas of physics. It was also important for the astronomical concept of the vacuum. If we ignored the possibility of Einstein's cosmological constant then it appeared that there could exist vacuum universes devoid of any ordinary matter. But if the cosmological constant is really a form of matter that is always present then there really are no true vacuum universes. The ethereal lambda energy is always there, acting on everything but remaining unaffected by the motion and presence of other matter.

Unfortunately, no one seems to have taken any notice of Lemaître's remark even though it was published in the foremost American science journal of the day. The early nuclear and elementary-particle physicists never found anything in their theories of matter that looked compellingly like the lambda stress. Its image amongst cosmologists ebbed and flowed. The Second World War intervened and changed the direction of physics towards nuclear processes and radio waves. Soon after it ended the interest of cosmologists was captured by the novel steady-state theory of the Universe first proposed by Fred Hoyle, Hermann Bondi and Thomas Gold. Like Friedmann's universes the steady-state universe expanded, but its density did not diminish with time. In fact, none of its gross properties changed with time. This steadiness was achieved by means of a hypothetical 'creation' process that produced new matter everywhere at a rate that exactly counterbalanced the dilution due to expansion. The rate required is imperceptibly small, just a few atoms appearing in each cubic metre every ten billion years. In contrast to the Big Bang[24] models, the steady-state theory had no apparent beginning when everything came into being at once. Its creation was continual.

At first, it looked as if this cosmological theory required a new theory of gravity to supersede Einstein's. It needed to include a new 'creation field' that could generate the steady trickle of new atoms and radiation needed to maintain the constant density of the Universe. In 1951, William McCrea,[25] a British astrophysicist, showed that nothing so radical was required. The creation field could be added into Einstein's equations as an extra source of energy and mass. And when it was, it looked just like the lambda term. No continual creation was needed.

Sadly for its enthusiastic inventors, the steady-state universe was soon consigned to the history books. It was a good scientific theory because it made very definite predictions: the Universe should look, on average, the same at all epochs. This made it extremely vulnerable to observational test. In the late 1950s, astronomers started to amass evidence that the Universe was not in a steady state. The population of galaxies of different sorts changed significantly over time. Quasars were discovered to populate the Universe more densely in the past than today. Finally, in 1965, the remnant heat radiation from a hot past Big Bang state was detected by radio astronomers and modern cosmology was born.

During the mid-1960s, when the first quasars were discovered with redshifts clustered around a single value, it was proposed that a large enough lambda stress might have been able to slow the expansion of the Universe temporarily in the past when it was about a third of its present extent. This could have led to a build-up of quasar formation close to this epoch. However, this idea faded away as more and more quasars were found with larger redshifts and it began to be appreciated how the apparent confinement of their redshifts to lie below a particular value was an artefact of the methods used to search for them.

Since that time observational astronomers have been searching for definitive evidence to determine whether the Universe is expanding fast enough to continue expanding for ever or whether it will one day reverse into contraction and head for a big crunch. If a lambda force exists that is large enough to dominate the attractive force of gravity over very large extragalactic distances, then it should affect the expansion of the Universe in the way shown in Figure 6.5. The most distant clusters of galaxies should be accelerating away from one another rather than continually decelerating as they expand.

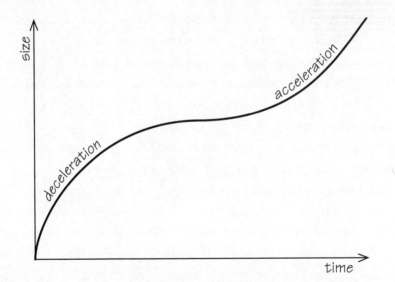

Figure 6.5 *The effect of lambda on the expansion of the Universe. When it becomes larger than the inverse square force of gravity it causes the expansion of the Universe to switch from deceleration to acceleration.*

The search for this tell-tale cosmic acceleration needs ways to measure the distances to faraway stars and galaxies. By looking at the change in the pattern of light colours coming from these objects we can easily determine how fast they are expanding away from us. This can now be done to an accuracy of a few parts in a million. But it is not so easy to figure out how far away they are. The basic method is to exploit the fact that the apparent brightness of a light source falls off as the inverse square of its distance away from you, just like the effect of gravity. So if you had a collection of identical 100-watt light bulbs located at different distances from you in the dark, then their *apparent* brightnesses would allow you to determine their distances from you, assuming there is no intervening obscuration. If you didn't know the intrinsic brightness of the bulbs, but knew that they were all the same, then by comparing their apparent brightnesses you could deduce their relative distances: nine times fainter means three times further away.

This is just what astronomers would like to be able to do. The trouble is Nature does not sprinkle the Universe with well-labelled identical light bulbs. How can we be sure that we are looking at a population of light sources that have the same intrinsic brightnesses so that we can use their apparent brightnesses to tell us their relative distances?

Astronomers try to locate populations of objects which are easily identifiable and which have very well-defined intrinsic properties. The archetypal example was that of variable stars which possessed a pattern of change in their brightness that was known theoretically to be linked to their intrinsic brightness in a simple way. Measure the varying light cycle, deduce the intrinsic brightness, measure the apparent brightness, deduce the distance away, measure the spectral light shift, deduce the speed of recession and *voilà*, you can trace the increase of speed with distance and see the expansion of the Universe, as Edwin Hubble first did in 1929 to confirm Friedmann's prediction from Einstein's theory that the Universe is expanding, as shown in Figure 6.6.

Unfortunately, these variable stars cannot be seen at great distances and ever since Hubble's work, the biggest problem of observational astronomy has been determining distances accurately. It is the twentieth-century

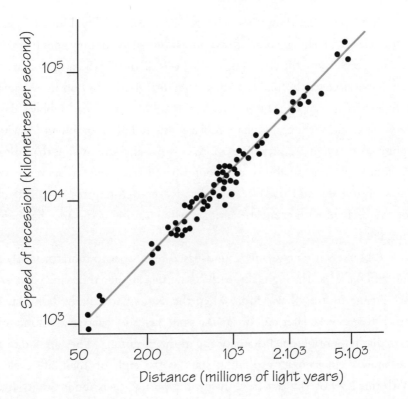

Figure 6.6 *Hubble's Law:*[26] *the increase of the speed of recession of distant sources of light versus their distance from us.*

analogue of John Harrison's[27] eighteenth-century quest to measure time accurately so that longitude could be determined precisely at sea. Until quite recently, it has made any attempt to map out the expansion of the Universe over the largest extragalactic dimensions too inaccurate to use as evidence for or against the existence of Einstein's lambda force. We could not say for sure whether or not the present-day expansion of the Universe is accelerating. Absence of evidence was taken as evidence of absence – and in any case it seemed to require a huge coincidence if the lambda force just started to accelerate the Universe at an epoch when human astronomers were appearing on the cosmic scene. Moreover, this would require lambda to have a fantastically small value. Better, they argued, to assume that it is really zero and keep on looking for a good reason why.

In the last year things have changed dramatically. The Hubble Space Telescope (HST) has revolutionised observational astronomy and, from its vantage point above the twinkling distortions of the Earth's atmosphere, it is now possible to see further than ever before. Telescopes on the ground have also advanced to achieve sensitivities undreamt of in Hubble's day. New electronic technologies have replaced the old photographic film with light recorders that are fifty times more sensitive at catching light than film. By combining the capability of ground-based telescopes to survey large parts of the sky and the HST's ability to see well-targeted, small, faint sources of light with exquisite clarity, a new measure of distance has been found.

Observers use powerful ground-based telescopes to monitor nearly a hundred pieces of the night sky, each containing about a thousand galaxies, at the time of the New Moon, when the sky is particularly dark. They return three weeks later and image the same fields of galaxies, looking for stars that have brightened dramatically in the meantime. They are looking for faraway supernovae: exploding stars at the ends of their life cycles. With this level of sky coverage, they will typically catch about twenty-five supernovae as they are brightening. Having found them, they follow up their search with detailed observations of the subsequent variation of the supernova light, watching the increase in the brightness to maximum and the ensuing fall-off back down to the level prior to explosion, as shown in Figure 6.7. Here, the wide sky coverage of ground-based telescopes can be augmented by the HST's ability to see faint light and colours.

The detailed mapping of the light variation of the supernovae enables the astronomers to check that these distant supernovae have the same light signature as ones nearby that are well understood. This family resemblance enables the observers to determine the relative distances of the distant supernovae with respect to the nearby ones from their apparent peak brightnesses, because their intrinsic brightnesses are roughly the same. Thus a powerful new method of determining the distances to the super-novae is added to the usual Doppler shift measurements of their spectra from which their speeds of recession are found. This gives a new and improved version of Hubble's law of expansion out to very great distances.

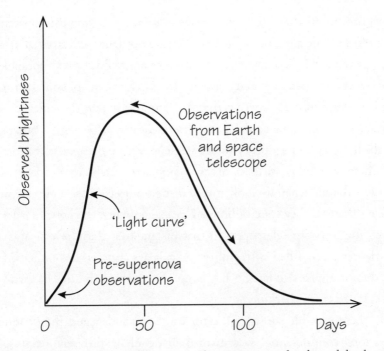

Figure 6.7 *A supernova light-curve. The variation in the observed brightness of a supernova, showing the characteristics to a maximum and gradual fall back to the level prior to the explosion.*

The result of these observations of forty distant supernovae by a combination of observations from the Earth and by the Hubble Space Telescope by two separate international teams of astronomers is to provide strong evidence that the expansion of the Universe is accelerating. The striking feature of the observations is that they require the existence of the cosmological constant, or lambda force. The probability that these observations could be accounted for by an expanding universe that is not accelerating is less than one in a hundred. The contribution of the vacuum energy to the expansion of the Universe is most likely[28] to be fifty per cent more than that of all the ordinary matter in the Universe.

The variation of the redshifting of light with distance for the sources of light cannot be made to agree with the pattern predicted if lambda does not exist. The only way of escape from lambda is to appeal to a mistake in the observations or the presence of an undetected astronomical process creating a bias in the observations, changing the apparent brightnesses of the supernovae so that they are not true indicators of distance, as assumed. These last two possibilities are still very real ones and the observers are probing every avenue to check where possible errors might have crept in. One worry is that the assumption that the supernovae are intrinsically the same as those we observe nearby is wrong.[29] Perhaps, when the light began its journey to our telescopes from these distant supernovae, there were other varieties of exploding star which are no longer in evidence. After all, when we look at very distant objects in the Universe we are seeing them as they were billions of years ago when the light first left them en route to our telescopes. In that distant past the Universe was a rather denser place, filled with embryonic galaxies and perhaps rather different than it appears today. So far, none of these possibilities has withstood detailed cross-checking.

If these possible sources of error can be excluded and the existing observations continue to be confirmed in detail by different teams of astronomers using different ways of analysing different data, as is so far the case, then they are telling us something very dramatic and unexpected: the expansion of the Universe is currently controlled by the lambda stress and it is accelerating. The implications of such a state of affairs for our understanding of the vacuum and its possible role in mediating deep connections between the nature of gravity and the other forces of Nature are very great.

So far we have seen what the astronomers thought about lambda and its possible role as the ubiquitous vacuum energy that Lemaître suggested. During the last seventy years, the study of the subatomic world has gathered pace and focus. It was also in search of the vacuum and its simplest possible contents. The discovery of a cosmic vacuum energy by astronomical telescopes turns out to have profound implications for that search too, and it is to this thread of the story that our attention now turns. It will start in the inner space of elementary particles and bring us, unexpectedly, full turn back to the outer space of stars and galaxies in our quest to understand the vacuum and its properties.

chapter seven

The Box That Can
Never Be Empty

"There is an element of tragedy in the life-story of the ether. First we had luminiferous ether. Its freely given services as midwife and nurse to the wave theory of light and to the concept of field were of incalculable value to science. But after its charges had grown to man's estate it was ruthlessly, even joyfully, cast aside, its faith betrayed and its last days embittered by ridicule and ignominy. Now that it is gone it still remains unsung. Let us here give it a decent burial, and on its tombstone let us inscribe a few appropriate lines:

Then we had the electromagnetic ether.
And now we haven't e(i)ther."

Banesh Hoffman[1]

IT'S A SMALL WORLD AFTER ALL

"This [quantum] theory reminds me a little of the system of delusions of an exceedingly intelligent paranoiac, concocted of incoherent elements of thoughts."

Albert Einstein[2]

One of the greatest truths about the character of the physical universe, which has come increasingly into the spotlight during the past twenty-five years, is the unity of its laws and patterns of change. Once upon a time it would have been suspected that the nature of the most elementary particles of matter had little to do with the shapes and sizes of the greatest clusters of galaxies in the astronomical realm. Likewise, few would have believed that a study of the largest structures in the Universe would be able to shed light upon the smallest. Yet, today, the study of the smallest particles of matter is inextricably linked to the quest for a cosmological understanding of the Universe and the organisation of matter within it. The reason is simple. The discovery that the Universe is expanding means that its past was hotter and denser than its present. As we retrace its history back to the first minutes, we encounter a cosmic environment of ever-increasing energy and temperature which ultimately reduces all the familiar forms of matter – atoms, ions and molecules – to their simplest and smallest ingredients. The number and nature of the most elementary particles of matter will thus play a key role in determining the quantities and qualities of the different forms of matter that survive the childhood of the Universe.

This cosmic link between the large and the small also features in the fate of the vacuum. We have just seen how the theory of gravity that Einstein created can be used to describe the overall evolution of the physical universe. In practice we choose a mathematically simple universe that is a very good approximation to the structure of the real one that we see through our telescopes. At first, we have seen how it was that Einstein's theory reinforced his expulsion of the ether from the vocabulary of physics by providing a natural mathematical description of universes which are completely devoid of mass and energy – 'vacuum' universes. No ether was necessary even if electrical and magnetic fields were introduced to curve space. Yet there was to be a sting in the tail of this new theory. It permitted a new force field to exist in Nature, counteracting or reinforcing the effects of gravity in a completely unsuspected way, *increasing* with distance so that it could be negligible in its terrestrial effects yet overwhelming on the cosmic scale of the Universe's expansion. This ubiquitous 'lambda' force allows itself to be interpreted as a new cosmic energy field: one that is omnipresent, preventing the realisation of nothing. But, if such a vacuum-buster exists, where does it come from and how is it linked to the properties of ordinary matter? Astronomers like Lemaître and McCrea posed these questions but did not answer them. They hoped that the world of subatomic physics would enable a link to be forged with the vacuum energy of Einstein.

Einstein's development of the theories of special and general relativity was one half of the story of the development of modern physics. The other half is the story of quantum physics, pioneered by Einstein, Max Planck, Erwin Schrödinger, Werner Heisenberg, Niels Bohr and Paul Dirac. Whereas the new theory of gravity was a single-handed creation by Einstein, needing no revision or interpretation, the quantum theory of the microworld was the work of many hands which had a tortuous path to clarity and utility. The task of unravelling what it meant combined challenging problems of mathematics with subtleties of interpretation and meaning, some of which are far from resolved even today. Each year several popular science books will appear which seek to explain the mysteries of quantum mechanics in a manner that readers who are not physicists will be able to understand.[3] Each of these authors is motivated to try yet another explanation of how it works by some of the unnerving words of warning from the founding fathers. From Niels Bohr, its principal architect,

> "Anyone who is not shocked by quantum theory has not understood it";[4]

or from Einstein,

> "The quantum theory gives me a feeling very much like yours. One really ought to be ashamed of its success, because it has been obtained in accordance with the Jesuit maxim: 'Let not thy left hand know what thy right hand doeth'";[5]

or Richard Feynman,

> "I think I can safely say that nobody understands quantum mechanics";[6]

or Werner Heisenberg,

> "Quantum theory provides us with a striking illustration of the fact that we can fully understand a connection though we can only speak of it in images and parables";[7]

or Hendrick Kramers,

> "The theory of quanta is similar to other victories in science; for some months you smile at it, and then for years you weep."[8]

Yet for all this ambivalence, the quantum theory is fabulously accurate in all its predictions about the workings of the atomic and subatomic worlds. Our computers and labour-saving electronic devices are built upon the things it has revealed to us about the workings of the microworld. Even the light-detectors that enable astronomers to see supernovae near the edge of the visible universe rely upon its strange properties.

The quantum picture of the world grew out of the conflicting pieces of evidence for the wavelike and particlelike behaviour of light. In some experiments it behaved as if it were composed of 'particles' possessing momentum and energy; in others it displayed some of the known properties of waves, like interference and diffraction. These schizophrenic behaviours were only explicable if energy possessed some revolutionary properties. First, energy is quantised: in atoms it does not take on all possible values but only a ladder of specific values whose separation is fixed by the value of a new constant of Nature, dubbed Planck's constant and represented by the letter h. An intuitive picture of how the wavelike character of the orbital behaviour leads to quantisation can be seen in Figure 7.1, where we can see how only a whole number of wave cycles can fit into an orbit.

Second, all particles possess a wavelike aspect. They behave as waves with a wavelength that is inversely proportional to their mass and velocity. When that quantum wavelength is much smaller than the physical size of

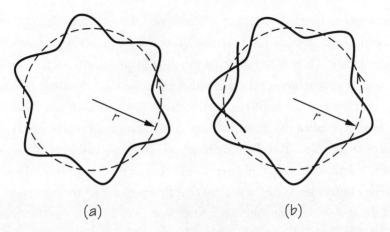

(a)   (b)

Figure 7.1 *Only a whole number of wavelengths will fit around a circular orbit, as in* (a) *but not in* (b).

the particle it will behave like a simple particle, but when its quantum wavelength becomes at least as large as the particle's size then wavelike quantum aspects will start to be significant and dominate the particle's behaviour, producing novel behaviour. Typically, as objects increase in mass, their quantum wavelengths shrink to become far smaller than their physical size, and they behave in a non-quantum or 'classical' way, like simple particles.

The wavelike aspect of particles turned out to be extremely subtle. The Austrian physicist Erwin Schrödinger proposed a simple equation to predict how a wavelike attribute of any particle changes in time and over space when subjected to forces or other influences. But Schrödinger did not have a clear idea of what this attribute was that his equation could so accurately calculate. Max Born was the physicist who saw what it must be. Curiously, Schrödinger's equation describes the change in the *probability* that we will obtain a particular result if we conduct an experiment. It is telling us something about what we can know about the world. Thus, when we say that a particle is behaving like a wave, we should not think of this wave as if it were a water wave or a sound wave. It is more appropriate to regard it as a wave of information or probability, like a crime wave or a wave of hysteria. For, if a wave of hysteria passes through a population, it means that we are more likely to find hysterical behaviour there. Likewise, if an electron wave passes through your laboratory it means that you are more likely to detect an electron there. There is complete determinism in quantum theory, but not at the level of appearances or the things that are measured. Schrödinger's amazing equation gives a completely deterministic description of the change of the quantity (called the 'wave function') which captures the wavelike aspect of a given situation. But the wave function is not observable. It allows you only to calculate the result of a measurement in terms of the probabilities of different outcomes. It might tell you that fifty per cent of the time you will find the atom to have one state, and fifty per cent of the time, another. And, remarkably, in the microscopic realm, this is exactly what the results of successive measurements tell you: not the same result every time but a pattern of outcomes in which some are more likely than others.

These simple ideas laid the foundations for a precise understanding of the behaviour of heat radiation and of all atoms and molecules. At first, they seem far removed from the definite picture of particle motion that Newton prescribed; but remarkably, if we consider the limiting situation where the particles are much larger than their quantum wavelengths, the quantum theory just reduces to the conclusion that the average values of the things we measure obey Newton's laws. Again, we see this important feature of effective scientific progress, that when a successful theory is superseded it is generally replaced by a new theory with an enlarged domain of applicability which reduces to the old theory in an appropriate limiting situation.

At first, the quantum theory seems to usher in a picture of a world that is founded upon chance and indeterminism, and indeed it was Einstein's belief that this was so, and it led him to spurn the theory he had helped to create as something that could not be part of the ultimate account of how things were. He could not believe that 'God plays dice'. Yet, on reflection, something like the quantum theory is needed for the stability of the world. If atoms were like little solar systems in which a single electron could orbit around a single proton with any possible energy, then that electron could reside at any radius at all. The slightest buffeting of the electron by light or distant magnetic fields would cause tiny shifts in its energy and its orbit to new values because all possible values are permitted. The result of this democratic state of affairs would be that every hydrogen atom (made of a proton and an electron) would be different: there would be no regularity and stability of matter. Even if all atoms of the same element started off identical, each atom in Nature would undergo its own succession of external influences which would cause a random drift in its size and energy. All would be different.

The quantum saves us from this. The electron can only occupy particular orbits around the proton, with fixed energies: hydrogen atoms can only have a small number of particular energies. In order to change the structure of the atom it must be hit by a whole quantum of energy. It cannot just drift into a new energy state that is arbitrarily close to the old one. Thus we see that the quantisation of atomic energies into a ladder of separate values, rather than allowing them to take on the entire continuum of possible values, lies at the heart of the life-supporting stability and uniformity of the world around us.

One of the most dramatic consequences of the wavelike character of all mass and energy is what it does for our idea of a vacuum. If matter is ultimately composed of tiny particles, like bullets, then we can say unambiguously whether the particle is in one half of a box or the other. In the case of a wave, the answer to the question 'Where is it?' is not so clear. The wave spans the whole box.

The first application of the quantum idea was made in 1900 by the great German physicist Max Planck, who sought to understand the way in which energy is distributed amongst photons of different wavelengths in a box of heat radiation – what is sometimes called 'black-body' radiation.[9] Observations showed that the heat energy apportioned itself over different wavelengths in a characteristic way. Our heat and daylight are provided by the Sun. Its surface behaves like a black-body radiator with a temperature of about 6000 degrees Kelvin.[10] There is little energy at short wavelengths. The peak is in the green part of the spectrum of visible light but most of the energy is emitted in the infrared region which we feel as heat (see Figure 7.2). The shapes of the curves change as the temperature increases in the manner shown in Figure 7.3. As the temperature increases, so more

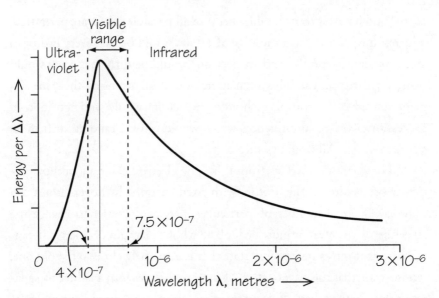

Figure 7.2 *The spectrum of a 'black-body' radiation source with a temperature of 6000 degrees Kelvin, similar to that of the Sun.*

energy is radiated at every wavelength, but the peak of the emission shifts to shorter wavelengths.

Tantalisingly, before Planck's work, it was not possible to explain the overall shape of this curve. The long-wavelength region where the energy steadily falls could be explained, as could the location of the peak, but not the fall towards short wavelengths. Planck was first able to 'explain' the curves by proposing a formula of a particular type. But this was not really *explaining* what was going on, merely describing it succinctly. Planck wanted a theory which predicted a formula like the one that fitted so well. He was impressed by the fact that the black-body energy distribution had a universal character. It did not matter what the emitter was made of; whether it was a flame, or a star, or a piece of hot iron, the same rule applied. Only the temperature mattered. It was a bit like Newton's law of gravity: the material out of which things are made does not seem to matter to gravity, they can be cabbages or kings; it is just their mass that determines their gravitational pull.

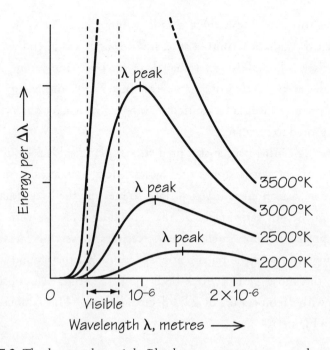

Figure 7.3 *The changing shape of the Planck curves as temperature, in degrees Kelvin, increases.*

Planck wanted to describe the behaviour of black-body radiation by the action of a collection of tiny oscillators, gaining energy by collisions with each other as heat is added, and losing energy by sending out electromagnetic waves at a frequency determined by that of the oscillations. Here, Planck had his most brilliant insight. In the past it had always been assumed that the oscillators in a system like this could emit *any* fraction of their energy, no matter how small. Planck proposed instead that the energy emission can only occur in particular quotas, or *quanta*, proportional to the frequency, *f*. Thus the energies emitted can only take the values *0, hf, 2hf, 3hf,* and so on, where *h* is a new constant of Nature,[11] which we now call Planck's constant. Planck modelled the whole glowing body as a collection of many of these quantised oscillators, each emitting light of the same frequency as it vibrates. Their energies can only change by a whole quantum step. At any moment it is more likely that the hot body contains more oscillators with low energies than those with high energies because the former are easier to excite. From these simple assumptions, Planck was able to show that the radiation emitted at each wavelength was given by a formula that precisely followed the experimental curves. The 'temperature' is a measure of the average value of the energy. Better still, the energy that his formula predicted would be emitted at wavelengths not yet observed subsequently proved to be correct.

Ever since these successful predictions, Planck's black-body law has been one of the cornerstones of physics. Most dramatically, in the last twelve years, astronomers have managed to measure the heat radiation left over from the hot early stages of the expanding Universe with unprecedented precision using satellite-based receivers, observing far above the interfering effects of the Earth's atmosphere. What they found was spectacular: the most perfect black-body heat spectrum ever observed in Nature, with a temperature of 2.73 degrees Kelvin.[12] This famous image is shown in Figure 7.4.

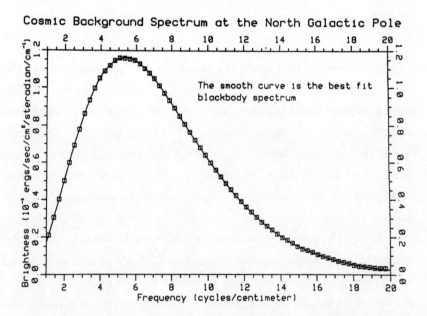

Figure 7.4 *The spectrum of the heat radiation left over from the early stages of the Universe and measured by NASA's Cosmic Background Explorer satellite (COBE). No deviations from a perfect Planck curve have been found.*

Planck's deductions about the nature of thermal equilibrium between matter and radiation at a given temperature were widely explored and ultimately led to the creation of a full quantum theory of all atomic interactions. This picture turned out to have one mysterious aspect to it. It described the intuitive idea of an equilibrium of radiation in a container. If the radiation started hotter than the walls of the container then the walls would absorb heat until they attained the same temperature as the radiation. Conversely, if the walls were initially hotter than the radiation they enclosed then they would emit energy that would be absorbed by the radiation until the temperatures were equalised. If you tried to set up an empty box whose walls possessed a finite temperature then the walls of the box would radiate particles to fill the vacuum.

As the implications of the quantum picture of matter were explored more fully, a further radically new consequence appeared that was to impinge upon the concept of the vacuum. Werner Heisenberg showed that there were complementary pairs of attributes of things which could not be measured simultaneously with arbitrary precision, even with perfect instruments. This restriction on measurement became known as the Uncertainty Principle. One pair of complementary attributes limited by the Uncertainty Principle is the combination of position and momentum. Thus we cannot know at once where something is *and* how it is moving with arbitrary precision. The uncertainty involved is only significant for very small things with a size comparable to their quantum wavelength. One way of seeing why such an uncertainty arises is to recognise that the act of making a measurement always disturbs the thing being measured in some way. This was always ignored in pre-quantum physics. Instead, the experimenter was treated like a bird-watcher in a perfect hide. In reality, the observer is part of the system as a whole and the perturbation created by an act of measurement (say light bouncing off a molecule and then being registered by a light detector) will change the system in some way. Another, more sophisticated and more accurate, way to view the Uncertainty Principle is as a limit on the applicability of classical notions like position and momentum in the description of a quantum state. It is not that the state has a definite position and momentum which we are prevented from ascertaining because we change its situation when we measure it. Rather, it is that classical concepts like position and velocity cannot coexist when one enters the quantum regime. In some ways this is not entirely surprising. It would be a very simple world if all the quantities that describe the behaviour of very big things were exactly those that were needed to describe very small things. The world would need to be the same all the way down to nothing.

THE NEW VACUUM

"The vacuum is that which is left in a vessel after we have removed everything which we can remove from it."

James Clerk Maxwell[13]

The Uncertainty Principle and the quantum theory revolutionised our conception of the vacuum. We can no longer sustain the simple idea that a vacuum is just an empty box. If we could say that there were no particles in a box, that it was completely empty of all mass and energy, then we would have to violate the Uncertainty Principle because we would require perfect information about motion at every point and about the energy of the system at a given instant of time. As physicists investigated more and more physical systems in the light of quantum theory, they found that the last stand mounted by the Uncertainty Principle manifested itself in the form of what became known as the *zero-point energy*. If one looked at the impact of quantisation on systems like the oscillators that lay at the heart of Planck's description of heat radiation equilibrium, it emerged that there was always a basic irreducible energy present that could never be removed. The system would not permit all its energy to be extracted by any possible cooling process governed by the known laws of physics. In the case of the oscillator, the zero level was equal to one-half of hf, the quantum of energy.[14] This limit respects and reflects the reality of the Uncertainty Principle in that if we know the location of a particle oscillator then its motion, and hence its energy, will be uncertain, and the amount is the zero-point motion.

This discovery at the heart of the quantum description of matter means that the concept of the vacuum must be somewhat realigned. It is no longer to be associated with the idea of the void and of nothingness or empty space. Rather, it is merely the emptiest possible state in the sense of the state that possesses the lowest possible energy: the state from which no further energy can be removed. We call this the *ground state* or the *vacuum state*.

As an illustration, consider a rather corrugated terrain of valleys and hills of different depths and heights, like that in Figure 7.5. The valley bottoms are the different minima of the system. They have different heights and are characterised locally by the simple fact that if you move slightly away from them in any direction you must travel uphill. One of these minima is lower than the others and is called the global minimum. The others are merely local minima. In the study of energies of systems of elementary particles of matter, such minima are called *vacua* to emphasise the characterisation of the vacuum by a minimum energy state. This example also illustrates something that will prove to have enormous importance for our understanding of the Universe and the structures within it: it is possible for there to be many different minimum energy states, and hence different vacua, in a given system of matter.

Indirect evidence for the physical reality of the zero-point energy appears every time a successful prediction emerges from quantum theories of the behaviour of radiation and matter. However, it is important to have

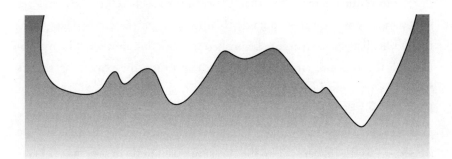

Figure 7.5 *An undulating terrain displaying several local peaks and valleys.*

a direct probe of its existence. The simplest way to do this was suggested by the Dutch physicist Hendrik Casimir in 1948 and has been known ever since as the Casimir Effect.

Casimir wanted to instigate a way for the sea of zero-point fluctuations to manifest themselves in an experiment. He came up with several ideas to achieve this, of which the simplest was to place two parallel, electrically conducting metal plates in the quantum vacuum. Ideally, the experiment should be performed at absolute zero temperature (or at least as close to it as it is possible to achieve). The plates are set up to reflect any black-body radiation that falls on them.

Before the plates are added to it we can think of the vacuum as a sea of zero-point waves of all wavelengths. The addition of the plates to the vacuum has an unusual effect upon the distribution of the zero-point waves. Only rather particular waves can exist between the two plates. These are waves which can fit in a whole number of undulations between the plates. The wave has to begin with a zero amplitude at the plate and end in the same way on the other plate. It is like attaching an elastic band between the two plates and setting it vibrating. It will be fixed at each end and the vibrations will undergo one, or two, or three, or four, or more, complete vibrations before the other plate is reached (see Figure 7.6).

The simple consequence of this is that those zero-point waves which do not fit an exact number of wavelengths between the plates cannot reside there, but there is nothing stopping them from inhabiting the region of space outside the plates. This means that there must be more zero-point fluctuations outside the plates than between them. Therefore the plates get hit by more waves on their outside than they do on the inside-facing surfaces. The plates will therefore be pushed towards one another. The magnitude of the pressure (force per unit area) pushing the plates together is $\pi hc/480d^4$, where d is the distance between the plates, c is the speed of light, and h is Planck's constant. This is called the Casimir Effect.[15] As you might expect, it is very small. The closer the plates can be placed (the smaller d) so the bigger will be the pressure pushing them together. This is to be expected since the effect arises because some wavelengths have been excluded from the collection between the plates as they don't fit in. If we

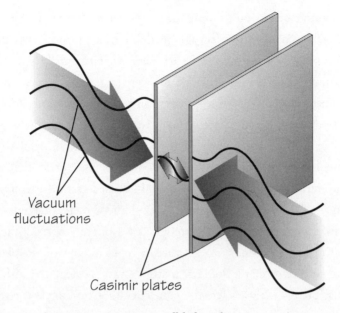

Vacuum
fluctuations

Casimir plates

Figure 7.6 *In the presence of a pair of parallel plates those vacuum energy waves that can fit a whole number of wavelengths between the plate will be present there. All possible wavelengths can still exist outside the plates.*

separate the plates a little further then more waves will be able to fit in and the disparity between the number of waves present between and outside the plates will get smaller. If the plates are separated by one half of one thousandth of a millimetre then the attractive pressure will be the same as that created by the weight of a fifth of a milligram[16] sitting on your finger tip, similar to that of a fly's wing.

Casimir had hoped that a spherical version of this model might provide a viable picture of the electron but unfortunately it was not possible to balance the repulsive electrostatic force against an attractive Casimir force as he expected. In fact, when one replaces the parallel plates in the zero-point sea by a spherical shell, or by other shapes, the calculations become very different (and very difficult) and the overall effect need not even have an attractive effect. The shape of the region placed in the vacuum is critically important in determining the magnitude and sense of the resulting vacuum effect.[17]

Casimir's beautifully simple idea has been observed in experiments. The first claim to see the effect was made by Marcus Sparnaay[18] in 1958, using two plates one centimetre square made of steel and chromium. However, the uncertainties in the final results were large enough to be consistent with no attractive effect being present, and it was not until 1996 that a completely unambiguous detection of the effect was made by Steve Lamoreaux[19] in Seattle with the help of his student Dev Sen. One of the greatest difficulties in performing these experiments is ensuring that the two plates are very accurately aligned parallel. In order to see an attractive effect between the plates which is as small as Casimir predicts, one must be able to control their separations to an accuracy of 1 micron over a distance of 1 centimetre. This job can be made easier by replacing one of the plates by a spherical surface so that it does not matter how it is orientated with respect to the flat plate – it always sees the same curvature. So long as the spherical surface is almost flat (or, at least, is not significantly curved over a distance equal to the distance between its surface and the flat plate – in Lamoreaux's experiment the separation was varied between 0.6 and 11 microns, whilst the radius of curvature of the curved surface was two metres) the expected attractive force can be recalculated to high accuracy. In the experiment, the force is measured by attaching one of the surfaces to the end of one arm of a torsion pendulum. Both surfaces are made of gold-coated quartz to maximise conductivity and robustness. The other end of the pendulum arm is placed between two conducting plates across which there is a voltage difference. A precise measurement of this voltage difference enables one to determine the electric force needed to overcome the attractive Casimir force between the plates and keep them at a fixed separation. The separation is measured with a laser interferometer[20] which is able to detect twisting of the pendulum to an accuracy of 0.01 of a micron (see Figure 7.7). The measured attraction of about 100 microdynes agrees with Casimir's prediction to an accuracy of five per cent.

What these beautiful experiments show is that there really is a base level of electromagnetic oscillation in space after everything removable has been removed. Moreover, this base level changes as the plate separations are changed and it exists between the plates and outside the plates at

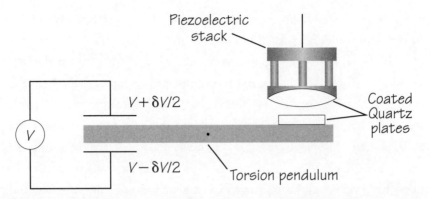

Figure 7.7 *The experimental set-up used to measure the Casimir force of attraction between two plates in a quantum vacuum.*

different levels. The energy in a given volume of the space between the plates is greater when the plates are closer than when they are far apart. This is understandable. If the plates attract one another you need to expend energy to separate them, after which the vacuum energy between them will be lower than before.

Even more ingenious experiments have been devised to probe the quantum fluctuations between the Casimir plates.[21] Atoms can be perturbed so that their electrons will change from one quantum orbital to another. When this happens they will emit light with a particular wavelength determined by the quantum of energy equal to the difference between the two energy levels. Allow this process to occur between a pair of Casimir plates and the normal decay will not be able to occur if the emitted light has a wavelength that does not fit between the plates. The atom will not decay as expected. Instead, it will remain in its perturbed state. If the wavelength of the emitted radiation fits nicely into the distance between the plates then the atom will decay more rapidly than it otherwise would in a space without the plates present.

There are many other experimentally observed effects of the zero-point energy. One of the earliest to be discovered was by Paul Debeye in 1914, who found that significant scattering of X-rays still occurred from the lattice of atoms making up a chunk of solid material even when the temperature started to approach absolute zero. This scattering is produced by the zero-point energy of the vibrations in the solid.

In the last few years a public controversy has arisen as to whether it is possible to extract and utilise the zero-point vacuum energy as a source of energy. A small group of physicists, led by American physicist Harold Puthoff,[22] have claimed that we can tap into the infinite sea of zero-point fluctuations. They have so far failed to convince others that the zero-point energy is available to us in any sense. This is a modern version of the old quest for a perpetual motion machine: a source of potentially unlimited, clean energy, at no cost.

While this more speculative programme was being argued about, wider interest in the vacuum was aroused by a phenomenon called 'sonoluminescence', which displays the spectacular conversion of sound-wave energy into light. If water is bombarded with intense sound waves, under the right conditions, then air bubbles can form which quickly contract and then suddenly disappear in a flash of light. The conventional explanation of what is being seen here is that a shock wave, a little sonic boom, is created inside the bubble, which dumps its energy, causing the interior to be quickly heated to flash point. But a more dramatic possibility, first suggested by the Nobel prize-winner Julian Schwinger,[23] has been entertained. Suppose the surface of the bubble is acting like a Casimir plate so that, as the bubble shrinks, it excludes more and more wavelengths of the zero-point fluctuations from existing within it. They can't simply disappear into nothing; energy must be conserved, so they deposit their energy into light. At present, experimenters are still unconvinced that this is what is really happening,[24] but it is remarkable that so fundamental a question about a highly visible phenomenon is still unresolved.

Puthoff (see note 22) has claimed far more speculative uses for vacuum energy, arguing that by manipulating zero-point energies we could reduce the inertia of masses in quantum experiments and open the way for huge improvements in rocket performance. The consensus is that things are far less spectacular. It is hard to see how we could usefully extract zero-point energy. It defines the minimum energy that an atom could possess. If we were able to extract some of it the atom would need to end up in an even lower energy state, which is simply not available.

ALL AT SEA IN THE VACUUM

"I must go down to the sea again, to the lonely sea and the sky,
And all I ask is a tall ship and a star to steer her by,
And the wheel's kick and the wind's song and the white sail's shaking,
And a grey mist on the sea's face and a grey dawn breaking."

John Masefield[25]

During the first half of the nineteenth century, an illustrated nautical book appeared in France[26] containing advice to mariners on how to deal with a host of dangerous situations encountered at sea. Some involved coping with adverse weather conditions and natural hazards, whilst others dealt with close encounters with other vessels. The Dutch physicist Sipko Boersma noticed that this handbook contained a peculiar warning to sailors of something that is reminiscent of the Casimir effect that we have just described.[27]

Sailors are warned that when there is no wind and a strong swell building, then two large sailing ships will start to roll. If they come close together and lie parallel to one another then they are at risk. An attractive force (*'une certaine force attractive'*) will draw the two ships together and there will be a disaster if their riggings collide and become entangled. The sailors are advised to put a small boat in the water and tow one of the ships out of the range of the attractive influence of the other. This sounds like a strange warning. Is there any truth to it? Remarkably, it turns out that there is. The attractive force between the ships arises in an analogous way to the force of attraction between the Casimir plates although there is no quantum physics or zero-point fluctuations of the vacuum involved – ships are too large for those effects to be big enough to worry about. Instead of waves of zero-point energy, the ships feel the pressure of the water waves.

The analogy is quite clear. Although we were dealing with radiation pressure between Casimir's plates, the same ideas apply to other waves as well, including water waves. In Figure 7.8, we see the situation of two ships, oscillating from side to side in the swell. The rolling ship absorbs energy from the waves and then re-emits this by creating a train of outgoing water waves. If the principal wavelength of these waves is much bigger than the distance between the two ships then they will rock together in time like a pair of copy-cat dancers. However, the waves that they radiate towards each other will be exactly out of phase. The peaks of one ship's waves will coincide with the troughs of the other ship's waves. The net result is that they will cancel each other out. As a result, there is virtually no radiated water-wave energy in between the two ships, and the pushing together of the ships, caused by the outgoing waves from the other sides of the ships, is not balanced. Thus, rolling ships will approach one another, just like atoms in a sea of vacuum fluctuations.

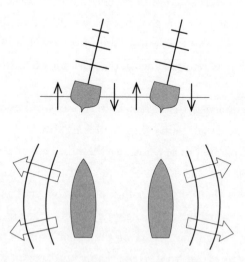

Figure 7.8 *Two nearby ships rolling in a swell of ocean waves. Some waves are excluded from the region between the ships and the ships are forced together by the higher wave pressure on their outer sides.*

The calculations[28] show that two 700-ton clipper ships should attract one another with a force equal to the weight of a 2000-kilogram mass. This is a reasonable answer. It is a force that a large boat of rowers could overcome by concerted effort. If the force were ten times bigger then all such efforts would be hopeless, whereas if it were ten times smaller the attraction would be negligible and no action would be needed to avert a collision. Boersma also discovered that the attractive force between the boats is proportional to the square of the maximum angle that they swing back and forth in the swell. In breezy conditions these oscillations will die out fairly quickly as the sails take up their energy. Thus we see the reason for the warning about the naufragous effects of coming too close to another ship in fairly calm conditions.

THE LAMB SHIFT

"I used to be Snow White . . . but I drifted."

Mae West[29]

One of the greatest successes of the quantum theory was to explain in exquisite detail the structure of atoms and the characteristic frequencies of the light waves that they emit when their electrons change from one quantum energy level to another. The first calculations of these levels got very accurate answers, in line with all observations, without realising that the vacuum energy might have an effect on the levels. Fortunately, the effect is very small and requires very sensitive measurements to detect it. It was not until 1947 that instruments were sensitive enough to detect these tiny changes. The electrons near the atomic nucleus feel tiny fluctuations created by the zero-point motions around them. These slight jigglings should result in a slight change in the path of the electron's orbit and a tiny shift of the energy level of the electron compared to its expected value if we ignore the vacuum fluctuations. In particular, in the hydrogen atom, two energy levels which would otherwise have the same level are split by a tiny amount, four millionths of an electron volt – more than three million times smaller than the energy needed to remove an electron from the atom. This tiny energy difference, now called the 'Lamb Shift', was first measured by the Americans Willis Lamb and Robert Retherford,[30] in 1947, using some of the techniques developed for the use of radar during the Second World War. Lamb received the Nobel prize for physics for this discovery in 1955.

FORCES OF THE WORLD UNITE

"God is in the atoms . . . A superposition, if you like. Or whether you don't like actually, that's what it's called. A superposition is like God in that the quantum object occupying a number of different states simultaneously can be everywhere at once. A superposition is a kind of immanence. Without these superpositions, quantum objects would simply crash into each other and solid matter could not possibly exist."

Philip Kerr[31]

The quantum vacuum with its seething mass of activity has ultimately proved to be the foundation of all our detailed understanding of the most elementary particles of matter. We have found only four distinct forces of Nature acting in the relatively low-energy world in which we live. Their properties are summarised in Figure 7.9. The action of each of these forces is sufficient to understand almost all the things that we see around us.[32] The quartet of forces includes gravity and electromagnetism, which are both familiar to us in everyday life, but they are joined by two microscopic forces which have only been explicitly isolated during the twentieth century. The 'weak' force lies at the root of radioactivity whilst the 'strong' force is responsible for nuclear reactions and the binding together of atomic nuclei. Each of these forces is described by the exchange of a 'carrier' particle which conveys the force. The quantum wavelength of this particle determines the range of influence of the force. The force of gravity is carried by

Force	Range	Relative Strength	Acts on	Carrier Particle
Gravity	Infinite	10^{-39}	Everything	Graviton
Electromagnetism	Infinite	10^{-2}	Electrically-charged particles	Photon
Weak	10^{-15} cm	10^{-5}	Leptons and hadrons	W & Z bosons
Strong	10^{-13} cm	1	Colour-charged particles	Gluons

Figure 7.9 *The four known fundamental forces of Nature.*

the exchange of a massless particle, the graviton, and so has an infinite range.[33] Gravity is unique in that it acts on every particle. The force of electromagnetism also has an infinite range because it is carried by the exchange of another massless particle, the photon of light. It acts on every particle that possesses electric charge. The weak interaction is different. It acts upon a class of elementary particles called leptons (Greek for light ones), like electrons, muons, tauons, and their associated neutrinos, and is carried by three very massive particles, the so-called intermediate vector bosons (W^+, W^- and Z^0). These particles are about 90 times heavier than the proton and the weak force they mediate has a finite range 100 times less than the radius of an atomic nucleus.

The strong force is more complicated. Originally, it was regarded as acting between particles like protons which undergo nuclear reactions. However, experiments in which these particles were collided at high energies revealed that they did not behave as if they were elementary indivisible pointlike particles at all. Rather, the proton deflected incoming particles as if it contained *three* internal pointlike scatterers. These internal constituents are known as quarks and they possess an analogue of electric charge that is called colour charge. This has nothing to do with the usual meaning of colour, as a hue determined by the wavelength of light absorbed when we observe it. It is just a particular attribute (like electric charge) which is conserved in all the processes that we have ever observed. The strong force acts on every particle that carries the colour charge and for this reason is sometimes called the 'colour force'. The colour force is mediated by the exchange of particles called gluons which have masses about 90 times less than the W and Z bosons and so the strong force has a range about 90 times greater. It is equal to the size of the largest atomic nucleus, a reflection of the fact that it is this force that binds it together.

Quarks possess both colour charge and electric charge. Gluons also possess colour charge and are therefore very different to photons. Photons mediate the electromagnetic interactions between electrically charged particles but do not themselves possess that electric charge – you can't have electromagnetic interactions of photons alone, they need charged particles like electrons to participate as well. Gluons, by contrast, carry the colour charge and mediate interactions between particles that possess colour charge – you could have strong interactions of gluons alone without any quarks. In this respect the gluons are more akin to the gravitons which mediate the gravitational force. Since gravity acts on everything that has mass or energy it also acts on the gravitons which convey it.

The most elementary particles of matter are believed to be the families of identical quarks and leptons listed in Figure 7.10. 'Elementary' means that they display no evidence of possessing internal structure or constituents.

The story of how this picture was established and the feats of engineering performed to establish the identities of the particles involved

First family	Second family	Third family
electron	muon	tau
electron neutrino	muon neutrino	tau neutrino
up quark	charmed quark	top quark
down quark	strange quark	bottom quark

Figure 7.10 *The three known 'families' of quarks and leptons. Each pair of quarks is related to a charged lepton (either the electron, muon or tau) and an uncharged neutrino.*

and the roles they play in Nature's great particle play is one about which whole books have been written.[34] Our interest is in a particular chapter of the story which reveals the reality and crucial properties of the quantum vacuum.

This theory appears succinct and appealing. It enables us to explain just about everything that is seen and has enabled a succession of successful predictions to be made. However, there is something unattractively incomplete about it all. Physicists believe deeply in the unity of Nature. A universe that rests upon four fundamental laws governing different populations of particles appears to them like a house divided against itself. The unity of Nature reveals itself in a host of different places and provokes us to show that these forces are not really different. If only we could find the right way of looking at them they would fall into place as different pieces of a single big picture, different parts of just one basic force of Nature from which everything derives. An analogy might be found in the behaviour of water. We see it in three very different forms: liquid water, ice and steam. Their properties are different yet they are all manifestations of a single underlying molecular structure for a combination of two hydrogen atoms and one oxygen atom. Despite appearances there is an underlying unity.

Any attempt to unify the quartet of basic forces seems doomed from the start. They look too different. They act upon different classes of elementary particles and they have very different strengths. The relative strengths are shown in Figure 7.9. We see that gravity is by far the weakest. The gravitational force between two protons is about 10^{38} times weaker than the electromagnetic force.[35] At laboratory energies, the weak force is about a hundred million times weaker than the electromagnetic force and the strong force is ten times stronger than electromagnetism.

The fact that the four separate forces have such different strengths and act upon separate sub-populations of elementary particles is deeply perplexing for anyone seeking a hidden unity behind the scenes that would unite them into a single superforce described by one all-encompassing 'theory of everything'. How can they be united when they are so different? The answer that has emerged reveals the vacuum to be the key player.

VACUUM POLARISATION

"Thirty spokes share the wheel's hub
It is the centre hole that makes it useful.
Shape clay into a vessel;
It is the space within that makes it useful.
Cut doors and windows for a room;
It is the holes which make it useful.
Therefore profit comes from what is there;
Usefulness from what is not there."

Lao-tzu[36]

We used to think of the strength of a force of Nature like electro-magnetism as a fixed constant of Nature, one of the defining features of the Universe. It could be described by combining the basic unit of electric charge carried by a single electron, the speed of light in a vacuum, and Planck's constant, h. These can be organised into a combination that possesses no units of mass, length, time or temperature. Thus, it provides us with a universal measure of the strength of electromagnetic forces of Nature irrespective of the units of measurement that we employ for the pieces that go into it (so long as we use the same units for all of them). The value obtained[37] by experiments of great accuracy and ingenuity for this pure number, called the *fine structure constant* and denoted by the Greek letter alpha, is equal to

$$\alpha = 1/137.035989561\ldots$$

Usually, it is regarded as being approximately equal to $1/137$ and physicists would love to explain why it has the precise numerical value that it does. We say that it is a fundamental constant of Nature. Accordingly, the number 137 is instantly recognised by physicists as significant and I have no doubt that the key codes of the locks on the briefcases of a significant number of physicists around the world involve the number 137. For an example of the type of numerological flights of fancy that this quest can inspire see Figure 7.11.

The fine structure constant tells the strength of the interaction that occurs when we fire two electrons towards each other. They have the same

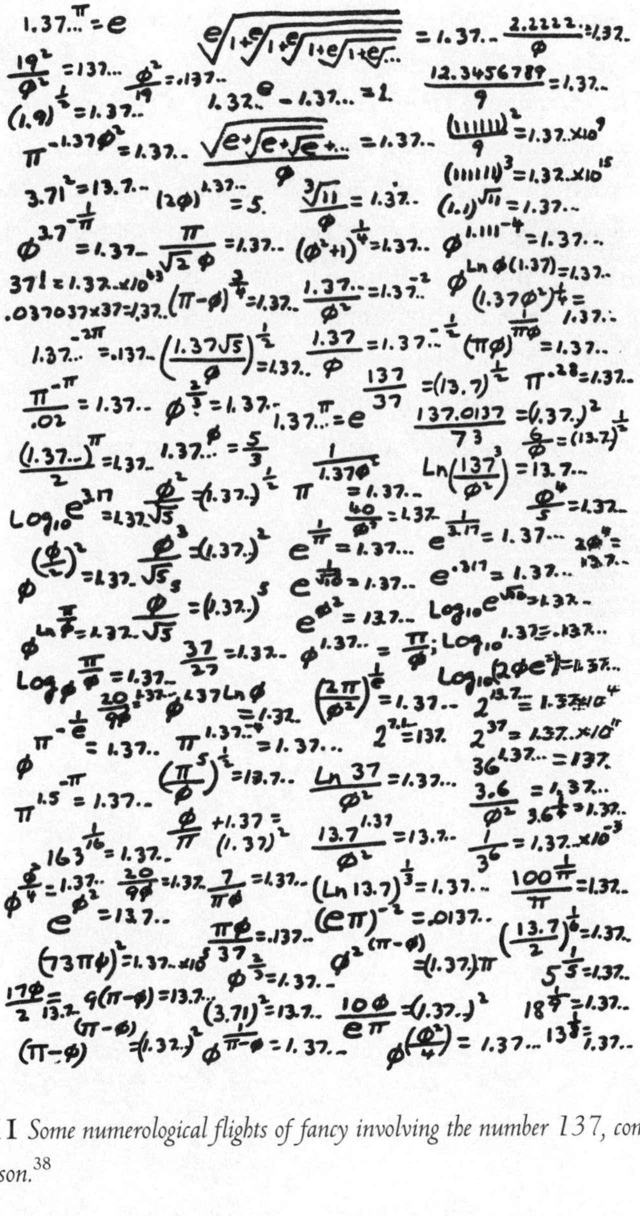

Figure 7.11 *Some numerological flights of fancy involving the number 137, compiled by Gary Adamson.*[38]

(negative) electric charge and so they repel one another like two magnetic North poles (see Figure 7.12).

In a world without quantum mechanics this interaction should produce the same degree of deflection regardless of the temperature or energy of the environment. All that counts is the number $1/137$. In a nineteenth-century vacuum composed of empty space there would be nothing more to be said.

The quantum vacuum changes all that. Our two electrons are no longer situated in a completely empty space – the Uncertainty Principle forbids us from entertaining any such notion. They are moving in the quantum vacuum and that is far from empty. It is a hive of activity. You recall that the Uncertainty Principle reveals that there are complementary pairs of properties that we cannot measure at once with unlimited precision. The energy and lifetime of a particle or a collection of particles is one of these so-called 'complementary' pairs. If you want to know everything about the energy of a particle you have to sacrifice all knowledge

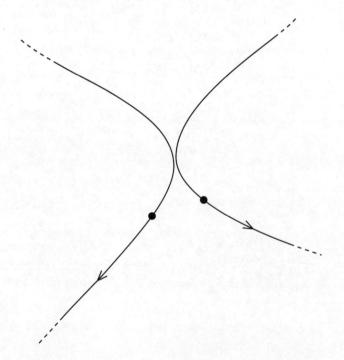

Figure 7.12 *Two electrons deflecting in a world with an empty 'classical' vacuum.*

about its lifetime. Heisenberg's Uncertainty Principle tells us that the product of these uncertainties always exceeds Planck's constant divide by twice the number pi:

(uncertainty in energy) $\times$ (uncertainty in lifetime) $> h/2\pi$ (*).

Any observed particle or physical state must obey this inequality. Observability requires that it be satisfied.

The quantum vacuum can be viewed as a sea composed of all the elementary particles and their antiparticles continually appearing and disappearing. For example, let us focus attention upon the electromagnetic interactions only for the moment. There will be a ferment of electrons and positrons.[39] Pairs of electrons and positrons will appear out of the quantum vacuum and then quickly annihilate each other and disappear. If the electron and the positron each have mass m, then Einstein's famous formula ($E = mc^2$) tells us their 'creation' requires an energy equal to $2mc^2$ to be borrowed from the vacuum. If the time they exist before annihilating back into the vacuum is so short that the Uncertainty Principle (*) is *not* obeyed, so

(uncertainty in energy) $\times$ (uncertainty in lifetime) $< h/2\pi$, (**)

then these electron-positron pairs will be *unobservable*. Hence, they are called *virtual* pairs. If they live long enough for (*) to be satisfied before they annihilate each other and disappear, then they will become observable and are called *real* pairs. The creation of virtual pairs seems like a violation of the conservation of energy. Nature allows you to violate this principle so long as no one can see you doing it and this is guaranteed so long as you repay the energy quickly enough. It is useful to think of the virtual condition (**) rather like an 'energy-loan' arrangement. The more energy you borrow from the energy bank the quicker you have to pay it back before it is noticed.

The upshot of this is that we can think of our quantum vacuum as containing a collection of continually appearing and disappearing virtual pairs of electrons and positrons. This sounds a little mystical, for if they are unobservable why not just ignore them and opt for a simpler life? But let us reintroduce our two electrons that are all set to interact. Their presence creates an important change in the quantum vacuum. Opposite electric charges attract and so if we put an electron down in the vacuum of virtual pairs the positively charged virtual positrons will be drawn towards the electron, as shown in Figure 7.13(a).

The electron has created a segregation of the virtual pairs and the electron finds itself surrounded by a cloud of positive charges. This process is called *vacuum polarisation*. Its effect is to create a positively charged shield around the bare negative charge of the electron. An approaching electron will not feel the full negative electric charge of the electron sitting in the vacuum. Rather, it will feel the weaker effect of the shielded charge and be scattered away more feebly than if the vacuum polarisation was absent.

This effect changes if we alter the energy of the environment and the incoming electron. If it comes in rather slowly, then it will not penetrate very far into the shielding cloud of positive charges and will be deflected

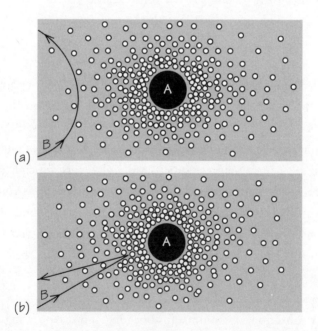

Figure 7.13 *(a) An incoming electron (B) with low energy scatters weakly due to the outer shield of virtual positive charges around the central negative charge of the electron (A); (b) an incoming electron with high energy penetrates the cloud of positive virtual charges and feels a strong repulsion from the central negative charge of the second electron.*

weakly. But, if it comes in with a higher energy, it will penetrate further through the shield and feel the effect of more of the full negative electron charge within. It will be deflected more strongly than the less energetic particle. Thus we see that the effective strength of the electromagnetic force of repulsion between the two electrons depends upon the energy at which it takes place, as shown in Figure 7.13(b). As the energy increases so the interaction appears to get stronger. It is a little like covering two hard billiard balls with a soft woolly padding. If the balls collide very gently then they will deflect only slightly because the hard surfaces will not hit and rebound. Only the woolly shields will gently rebound. But if they are made to collide at high speed the shields will have little effect and the balls will rebound very strongly. The trend is clear: as the energy of the environment increases so the stronger does the *effective* electromagnetic interaction become. As the energy rises, the incoming particle gets a closer 'look' at the bare point electron charge beneath the cloud of virtual positrons and is deflected more.

The same study can be made of the strong interaction that affects particles, like quarks and gluons, which carry the colour charge. The situation is a little more complicated than that of the electromagnetic interaction. When we considered the effects of the repelling charges of virtual electrons and positrons we could ignore the photons mediating their electromagnetic interaction because they have no electric charge. However, if we put a quark of fixed colour charge down in the vacuum and fire another coloured quark towards it, there are two vacuum polarisation effects to consider. Just as before, there will be a cloud of quark-antiquark pairs which will tend to surround any quark with a screening cloud of opposite colour charge. As with the electrons, the overall effect will be to make the strong interaction effectively stronger at higher energies. However, the presence of the gluons also affects the pattern of colour charge. Virtual gluons have the opposite effect and tend to smear out the central colour charge. When scattering occurs from a more extended, less pointlike object it tends to be weaker. The winner between these two opposed tendencies depends on how many varieties of quark there are to pop up in virtual pairs. If the number is as low as the six that we observe in Nature, then it is the gluon smearing that wins out and the strong interaction is predicted to get effectively *weaker* as we go to higher energies.

This property, called '*asymptotic freedom*' because it implies that if one continues to extrapolate to indefinitely increasing energies there would be no apparent interaction at all – the particles would be free – was predicted in 1973 and was quite unexpected. It is now confirmed by observations of the interaction strength at different energies. It revolutionised the study of elementary particles and high-energy physics and opened the door to making serious studies of the first moments of the expanding Universe when temperatures would have been high enough for these effects to be very significant. Before 1973, there had been widespread belief that the strong interaction was going to be hopelessly complicated and there was not much chance of understanding interactions at very high energies. It was assumed that they got stronger and stronger at higher and higher energies and so became increasingly intractable. Asymptotic freedom meant that in many ways things got simpler and simpler and it was possible to make real progress.

These important effects of the quantum vacuum enable us to see how the puzzling obstacle to unification of the forces of Nature posed by their different apparent strengths might be overcome. The force strengths do indeed differ significantly in the low-energy world where life like ours is possible, but if we follow the changes expected in those forces as we go to higher and higher energies, they can become closer and closer in strength until a particular energy is reached where the strengths are the same (Figure 7.14). Unification exists only in the ultra-high-energy environment that would have existed in the early stages of the Universe. Today, things have cooled off, and we are left searching for the remnants of a symmetrical past, disguised by billions of years of history. At the energies of our life-supporting environment these forces look very different and the unity of the forces of Nature is hidden. The deep symmetry of the forces that should be found at high energies is possible only because of the contributions of the quantum vacuum. This sea of virtual particles is really there. Its effects can be observed, as predicted, by the change in strength of natural forces as energies increase. The vacuum is far from empty. Nor is it

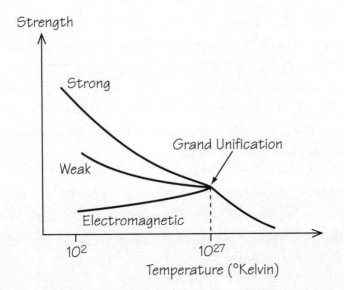

Figure 7.14 *Asymptotic freedom. The weakening of the strong force between quarks as the energy of interaction increases predicts that it can have the same strength as the electromagnetic force at very high temperatures.*

inert. Its presence can be felt and measured in the elementary-particle world, and without its powerful contribution, the unity of Nature could not be sustained.

BLACK HOLES

"Confinement to the Black Hole . . . to be reserved for cases of Drunkenness, Riot, Violence, or Insolence to Superiors."

British Army Regulation, 1844

One of the most recurrently fascinating concepts in the whole of science has proved to be that of the 'black hole'.[40] This cosmic cookie monster, relentlessly devouring everything that strays too close, has captured the popular imagination like no other scientific concept, starred in Hollywood movies, and inspired a host of science-fiction stories. Black holes are regions where the gravitational field of matter is too strong for anything, even light, to escape from its grasp. In Einstein's picture of curved space, the concentration of mass within a small region can grow so great that the geometry curves dramatically and pinches off the region surrounding it, preventing any signals getting out. The mass concentration is surrounded by a surface of no-return, called the *event horizon*, through which material and light can flow in but not out.

Despite their popular image, black holes are not necessarily solid objects possessing enormous densities. The huge black holes that seem to be lurking at the centres of many large galaxies have masses about a billion times larger than that of the Sun, but their average density is only about that of air.[41] We could be passing through the event horizon of one of these vast black holes at this very moment and nothing would seem strange. No alarms bells would ring as we crossed the horizon and we wouldn't be torn apart.[42] Later on, it would: gradually we would find ourselves drawn inexorably towards conditions of increasing density at the centre. If we ever tried to reverse our path, we would find that there was a definite limit to how far back we could get and none of the signals beamed back to base outside the hole would ever be received.

Black holes are predicted to form whenever a star that is more than about three times the mass of our Sun exhausts its nuclear fuel. It will then cease to have any means of supporting itself against the inward pull of gravity exerted by its constituents. No known force of Nature is strong enough to resist this catastrophic implosion and it will continue to compress the material of the star into a smaller and smaller region until a horizon surface is created. From the inside, the compression just carries on going but from the outside it ceases to be visible. A distant observer looking at the black hole would see light from just outside the horizon becoming redder and redder as it loses energy climbing out of the very strong gravity field.[43] The only trace that remains is its gravitational pull.

Evidence has steadily mounted to such an extent that the existence of black holes is regarded as established beyond all reasonable doubt by astronomers. The trick is to catch a black hole in orbit around another luminous star.[44] The orbit of the visible star will betray the presence of an unseen companion and the star will have material steadily pulled from its outer regions by the companion's gravitational pull. This material will be heated to millions of degrees Kelvin as it swirls down into the plughole produced by the black hole. At these temperatures there will be a profuse emission of X-rays from the heated material, colliding with other particles, on its inspiralling trajectory. When it nears the horizon surface the wavelength of the flickering of these X-rays will tell us the size of surface they are disappearing into. Black holes have a very particular relationship between their mass and the size of their event horizon. The information obtained from the motion of the visible star and the flickering of the X-rays enables this relationship to be checked. A number of such 'X-ray binary systems' are now known and they provide very strong evidence that black holes result when very massive stars end their careers and collapse in upon themselves.

Up until 1975 this picture of black holes was regarded as the full story. Things went into black holes. They never came out. But then the picture changed in a dramatic way. Stephen Hawking[45] asked what would happen if a black hole was placed in a quantum vacuum. Remember what we have just seen when the Casimir plates are placed in a quantum vacuum. The sea of vacuum fluctuations of all wavelengths is affected. Now imagine what would happen if a black hole were introduced. If a virtual particle-antiparticle pair appeared very close to the horizon then one of the particles could fall inside the horizon surface while the other stayed outside. The virtual particles would become real; the outgoing particle would be detected by a distant observer and the black hole would appear to be radiating particles from all over its horizon surface.[46] This process should be happening continuously and the net result is that all black holes will slowly evaporate away. Black holes are not truly 'black' when the quantum vacuum is taken into account. Further investigation revealed that the radiation of vacuum particles followed the laws of black-body thermodynamics originally discovered by Planck. Black holes were black bodies. Sadly, the rate at which particles are expected to be radiated is very slow when black holes are as large as those seen in X-ray binary star systems. In order for Hawking's radiation process to be visible,[47] we would have to encounter black holes which are only about the mass of a large mountain or asteroid. Their horizon size is equal to that of a single proton! These 'mini' black holes cannot form today when stars die. But they can be formed in the dense environment of the Big Bang if it is irregular enough. If they were, then these mountain-sized black holes would be in the final stages of evaporation today. The climax of the process will be a dramatic explosion that would show up as a burst of high-energy gamma rays accompanied by radio waves arising from the fast-moving electrons emerging from the explosion at speeds close to that of light. They would radiate 10 gigawatts of gamma-ray power for a period of more than forty billion years and could be seen many light years away. Radio telescopes could see the radio waves from one of these atomic-sized explosions occurring two million light years away in the Andromeda galaxy.

Observers have searched for evidence of black-hole explosions but none has yet been found. All we can say is that if exploding mini black holes do exist then they are few and far between, with no more than one occurring per year in every sphere of space, one light year in diameter.

The Hawking radiation process is of great significance for our understanding of the way in which the great laws underlying Nature are interwoven. It is a unique example of a process which is both relativistic, quantum, gravitational, and thermodynamic. Again, we see that its existence is a direct consequence of the reality of the vacuum and the sea of fluctuations within it. The steep gradient in the gravitational force field near the horizon of the black hole pulls the virtual pairs apart and prevents them annihilating back into the vacuum. They become real particles at the expense of the energy of the gravitational field of the black hole.[48]

In this chapter we have seen the vacuum move to centre stage in our story. Its existence and universality turn out to underlie the workings of all the forces of Nature. It influences the strengths of the electromagnetic, weak and strong forces of Nature, and links the force of gravity to the quantum character of energy. Each of these influences provides us with observational evidence for the reality of the quantum vacuum and the fluctuations that support it. These successes have flowed from a new conception of the vacuum that gives up the ancient picture of the vacuum as completely empty space. In its place is the more modest view that the vacuum is what is left when everything is removed from space that can be removed. What is left is the lowest energy state available. Remarkably, this means that the vacuum might change, steadily or suddenly. If it does then it can alter the complexion of the entire Universe. In the next chapter we see how.

chapter eight

How Many Vacuums Are There?

"Why is there only one Monopolies Commission?"

Screaming Lord Sutch [1]

VACUUM LANDSCAPE APPRECIATION

"The Grand Old Duke of York
He had ten thousand men,
He marched them up to the top of the hill,
And he marched them down again."

Nursery Rhyme

The subtleties and unexpected properties of the quantum vacuum elevated it to play a leading role in fundamental physics in the mid-1970s. Since then its position has become increasingly wide-ranging and pivotal. Every day sees new research papers about some aspect of the vacuum posted on the electronic web sites that physicists use to announce their new work to colleagues all over the world.[2] What has given rise to this explosion of interest? The adoption of a definition of the vacuum that requires it to be only a state of minimum energy is the answer. It immediately opens up a number of extraordinary possibilities.

The first question that we might pose about the vacuum as minimum energy state is, 'Why should there be only one of these minimum energy states?' The energy 'landscape' could contain many undulations, valleys and hills, just like a real terrain. These undulations could be very regular, like a corrugated roof or an egg box, with many different minima, each having the same minimum value for the energy (see Figure 8.1). This scenario suggests two new possibilities: if there can be many vacuums then we have to decide in which one of them our Universe is going to end up; also, we would like to know if it is possible to change vacuums in some way, by jumping from one minimum to the other.

In the example we have drawn in Figure 8.1, the different vacuums correspond to minima of the same depth. We could add a further dimension of possible variation to the situation by marking the position of the vacuum on a two-dimensional surface and its depth by the height above it. This is like a real landscape on the Earth's surface in which the height above or below sea level defines the altitude at each location. When this extra dimension is added it becomes possible for a continuous line of points to be vacuums at the same height for the system. A simple example is shown in Figure 8.2, where the vacuums form a ring on the floor. In the middle of the ring is a maximum so that the overall shape of the energy landscape is rather like a Mexican hat.

We can imagine still more unusual situations. We have drawn all the minima to lie at the same levels but there is no need for this. The vacuums are just defined by the presence of a local minimum in the landscape. There is no reason why they all need to be at the same level. If there is one which has a lower energy value than the others we will call it the 'true' or

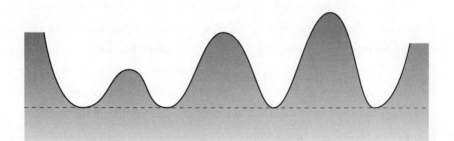

Figure 8.1 *A vacuum landscape with many local minima of equal depth.*

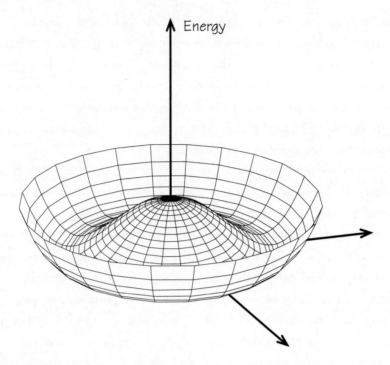

Energy

Figure 8.2 *A continuous circle of minima of the same depth.*

'global' vacuum. Also, the minima can differ in other, more subtle respects. The curvature of the terrain in their immediate vicinities can be different (see Figure 8.3). So, the terrain can rise steeply or gradually as we move away from the minimum. If you find yourself in a steep-sided vacuum it will be harder to escape compared than from the shallow-sided sort.

When we looked at some of the effects of vacuum polarisation on the strengths of the measured forces of Nature in the last chapter we saw how the temperature of the environment in which forces act matters. Thus we might well expect that our energy landscapes depend on temperature. As the temperature changes, the shape of the landscape can change very significantly. Both the number of vacuums may change as well as their depths. Some can even cease to be minima if the landscape changes very dramatically.

An interesting example of this process is provided by magnetism. The magnetisation energy of a bar of iron has a pattern of variation that is strongly dependent upon the temperature of the metal. When an iron bar

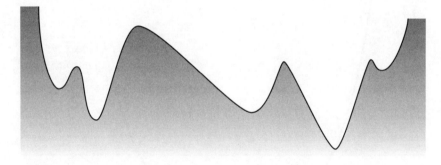

Figure 8.3 *Landscape with different minima and different gradients.*

is heated above a particular temperature of 750 degrees Celsius, called the Curie temperature, it displays no magnetic properties. There is no North and South magnetic pole on the bar. The high temperature has randomised the directions of all the atomic configurations in the iron and so there is no overall directionality to the bar's properties. As the bar is allowed to cool below the Curie temperature a spontaneous magnetisation takes place: the bar ends up with a North magnetic pole at one end and a South magnetic pole at the other. If you repeat this heating and cooling process a number of times you will not necessarily find that the North magnetic pole always lies at the same end of the magnet. We can understand what is happening by looking at the energy landscape above and below the Curie temperature as shown in Figure 8.4.

Above the Curie temperature, there is a single minimum vacuum state for the bar. It is symmetrically placed with the minimum at zero so there is no preference for one direction (right) of the bar over another (left). The minimum is a steep-sided valley into which everything will roll no matter where it starts out up the valley and this tells us that it doesn't matter how our piece of iron started out. Once it is hot enough it will enter this minimum unmagnetised state and lose memory of any previous magnetised state. However, as the bar cools below the Curie temperature something unusual happens. The magnetisation-energy landscape changes from having a single central valley into one with two valleys and a peak in between. The original minimum has turned into a precarious maximum, whilst two

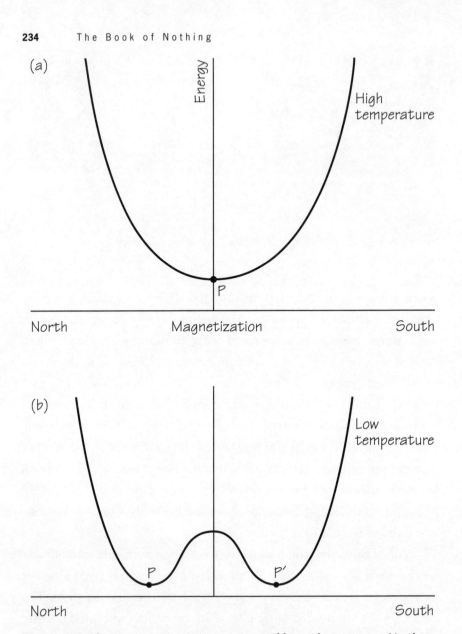

Figure 8.4 *The variation of magnetisation of a metal bar with temperature. (a) Above a critical temperature there is a single stable minimum (P) with no preferred directions. (b) Below the critical temperature two minima of equal depth appear and the previous stable minimum turns into an unstable maximum. A point located there will eventually roll into one of the two asymmetrical minima (P and P′) and the bar will have a magnetic North pole at one end and a magnetic South pole at the other.*

new deeper minima have appeared, equidistant on either side of the central maximum. What does this mean for our iron bar? It means that the symmetrical unmagnetised state has become unstable. The system will roll off down into one of the two new minima. There is an equal chance of going either way and this corresponds to the bar being magnetised with the right-hand end or the left-hand end as the magnetic North pole of the resulting bar magnet. This transition from a state where the minimum that the system resides in is symmetric about the zero value to one in which it is asymmetrical is a common phenomenon in Nature and it is called *symmetry breaking*.

The phenomenon of symmetry breaking reveals something deeply significant about the workings of the Universe. The laws of Nature are unerringly symmetrical. They do not have preferences for particular times, places and directions. Indeed, we have found that one of the most powerful guides to their forms is precisely such a requirement. Einstein was the first to recognise how this principle had been used only partially by Galileo and Newton. He elevated it to a central requirement for the laws of Nature to satisfy: that they appear the same to all observers in the Universe, no matter how they are moving or where they are located. There cannot be privileged observers for whom everything looks simpler than it does for others. To countenance such observers would be the ultimate anti-Copernican perspective on the Universe.[3] This democratic principle is a powerful guide to arriving at the most general expression of Nature's laws. Yet, despite the symmetry of the laws of Nature, we observe the *outcomes* of those symmetrical laws to be asymmetrical states and structures. Each of us is a complicated asymmetrical outcome of the laws of electromagnetism and gravity. We occupy particular positions in the Universe at this moment of time even though the laws of gravity and electromagnetism are completely democratic with respect to positions in space. One of Nature's deep secrets is the fact that the outcomes of the laws of Nature do not have to possess the same symmetries as the laws themselves. The outcomes are far more complicated, and far less symmetrical, than the laws. Consequently, they are far more difficult to understand. In this way it is possible to have a Universe governed by a very small number of simple symmetrical laws (perhaps just a single law) yet manifesting a stupendous array of complex, asymmetrical states and structures that might even be able to think about themselves. In the last decade, there has been an enormous upsurge of interest in trying to understand the asymmetrical outcomes of symmetrical laws. The availability of inexpensive fast computers has greatly facilitated this activity because the complexities of the asymmetrical outcomes are generally too great for unaided human calculation to reveal what is happening in full detail.

THE UNIFICATION ROAD

"Encyclopaedia Britannica full set, no longer needed due to husband knowing everything."

Personal ad, *Lancashire Post*[4]

The joining together of the forces of Nature is made possible by the variation in their strengths as the temperature rises. This process sees a coming together first of the electromagnetic and weak forces to create a single electroweak force when temperatures reach about 10^{15} degrees Kelvin. If we carry on charting the strengthening of this force together with the weakening of the strong force, then a second unification is implied when temperatures reach a level of about 10^{27} degrees Kelvin. Above this so-called 'grand unification' temperature there is a single symmetrical force, but below it there is a breaking of this symmetry to create the different strong and electroweak forces.[5]

This change of symmetry as the temperature falls will be reflected in the behaviour of all the material in the Universe during its very early stages. We can imagine the Universe expanding away from a Big Bang where the initial temperatures and energies are high enough to maintain complete unification of the strengths of the strong and electroweak forces. As the temperature falls below a particular value, these forces separate and go their different ways.

This perspective upon the change of forces during the very early Universe focused the attention of high-energy physicists and cosmologists upon some of the unusual things that might happen if these changes occurred in special ways. In particular, if the elementary particles in the Universe underwent a change of vacuum state, from a high to a low level, then it could make the whole Universe behave in novel and very attractive ways.

In gradually exploring the ramifications of these ideas for the Universe, interest has focused upon the consequences of a hypothetical type of matter existing in the early Universe. In order to avoid being too specific we call this a 'scalar' field. This means that at any point of space, and at any time, this field has only one attribute – its magnitude or intensity (a 'scale'). For example, the density of printer's ink on this page is a scalar field. The temperature in a room is a scalar field. But wind velocity is not a scalar because it is determined by a magnitude *and* a direction at every point and moment of time.

In the earliest stages of the Universe's history the temperature will be very much higher than today and we could expect new forms of matter to be formed which possess a diverse range of vacuum landscapes. Let's pick on one of these energy fields. This field could have any number of vacuum states of different levels. It need not correspond exactly to any field that we can observe today because it could have decayed away into radiation and other particles during the early stages of the Universe but, ultimately, our unified theory of all the forces of Nature should tell us what it is. Fields like this will possess two types of energy: a kinetic part associated with their motion, and a potential energy associated with their location. A simple analogy is provided by a swinging clock pendulum. When the bob is swinging through its lowest point it is moving at its fastest and its energy is entirely kinetic. As it rises up to its highest point it gradually slows down: its kinetic energy is transformed into potential energy as the bob works to overcome the downward force of gravity. Momentarily, when it stops at its highest point, before beginning its downward motion, its energy is entirely potential.

Energy fields in the early Universe can behave like the pendulum. When the kinetic part of the energy is the largest, the field will change very quickly, but when the potential energy is largest it will change very slowly. Now suppose that the types of changes in the potential shape that we have just been looking at could come into play during the first moments of the Universe's expansion. The scalar field could begin at high temperatures in a single stable vacuum state like that shown in Figure 8.5, but when the temperature falls below a particular value, T_c, a new vacuum state could appear at much lower energy.

What will happen? If the original vacuum state has a rather shallow gradient around it, then it is possible for the field to respond to all the buffetings and exchanges of energy with other particles and radiation by jumping over the hill and moving off down towards the new minimum. If the transition takes place slowly enough the potential energy of the slowly moving field will hardly be diluted by the expansion of the Universe that is going on all around it. Meanwhile, all the other radiation and energy in the Universe is being rapidly diluted by the expansion and, consequently, the influence of the scalar field can quickly overwhelm everything else and be the dominant form of mass and energy in the Universe. If that happens, there are many dramatic consequences. The expansion of the Universe

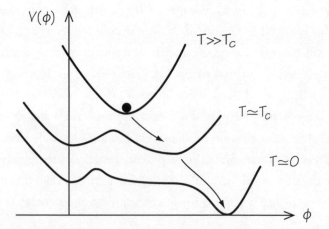

Figure 8.5 *The appearance of a new minimum.*

changes from steady deceleration to *acceleration*. This new state of affairs arises because the slowly changing scalar field behaves as if it is gravitationally *repulsive* whereas other forms of matter and radiation are invariably gravitationally attractive. This acceleration will continue for as long as the field rolls very slowly down the potential landscape. Whilst this slow change occurs, the acceleration will produce a very fast fall-off in the radiation temperature of the Universe. Eventually, the acceleration will come to an end. When the scalar field reaches the new vacuum state it will oscillate backwards and forwards many times, gradually losing energy, and decaying into other particles. Huge amounts of energy will be released from these decays and the temperature fall-off of the Universe created by the expansion will be dramatically slowed. The expansion will resume its normal decelerating course (see Figure 8.6).

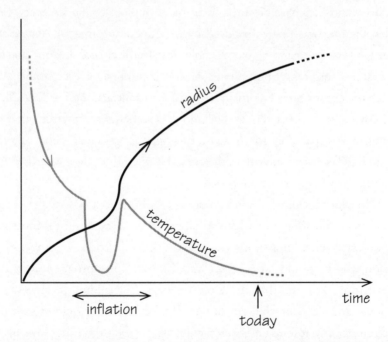

Figure 8.6 *The surge in expansion and fall in temperature created by a period of inflation in the early Universe. When inflation ends there is a complicated sequence of events, involving the decay of the scalar field driving the inflation, and the Universe heats up. Subsequently, it cools steadily and continues to expand at a slower rate.*

The hypothetical sequence of events we have just traced describes what has become known as cosmological 'inflation'. Inflation is an interval of cosmic history during which the expansion *accelerates*. It arises whenever a matter field, like a scalar field, changes very slowly from one vacuum state to another. In fact, it can also occur if there is only one vacuum state, so the potential landscape looks like a very shallow 'U' shape. The Russian physicist Andrei Linde,[6] now based at Stanford, California, pointed out that as the Universe cools down, a scalar field may just find itself starting to roll down the slope from an energy level high up the hill. If the slope is shallow enough the scalar field will change its energy so slowly that the energy of motion will always be negligible and anti-gravitation and inflation will arise. As physicists started to explore all the different ways in which this phenomenon could occur, it seemed that it was very difficult to avoid it.

The cosmological consequences of changing vacuums are rather extraordinary and they have been the focus of cosmologists' interest since 1981 when the idea was first introduced by the American physicist Alan Guth.[7] Our Universe is expanding tantalisingly close to the critical dividing line that separates a future in which the expansion continues for ever from one in which the expansion is eventually reversed into contraction. The 'critical', or in-between, universe is very special and it is somewhat mysterious that our Universe is expanding so close to this special trajectory. The universes expanding faster or slower than the critical case tend to diverge further away from the dividing line as time goes on.

In order for our Universe to be still within about twenty per cent of the critical rate after nearly fifteen billion years of expansion it must have begun expanding fantastically close to the critical divide. We know of no reason why it should have begun like that. Inflation offers an appealing explanation. Imagine that the Universe begins expanding in any way we choose, far away from the critical rate. If a scalar matter field exists which ends up rolling towards a lower vacuum state, then the expansion of the Universe will accelerate. For as long as it does, the expansion will be driven very rapidly, closer and closer towards the critical dividing line.

In this way, a very brief interval of inflation is sufficient to drive the expansion so close to the critical divide by the time inflation ends that the subsequent non-inflationary expansion will have a negligible effect on our distance from the critical divide, and we will find ourselves observing a universe that is expanding at a rate within about one part in 100,000 of the critical value.

This is not all. Another mystery of our Universe is the way in which its expansion rate is the same in every direction and from place to place with remarkable precision. If we scan the radiation reaching us from the edge of the visible Universe, we find that its temperature and intensity is the same in every direction to an accuracy of about one part in 100,000. Yet, as we run the history of the Universe backwards, this becomes very hard to understand. Light has not had time to cross from one side of the Universe to the other. There has not been time for differences in the temperature and density of the Universe from one place to another to have been ironed out in the time apparently available. However, if inflation occurred early on, the ensuing surge of accelerated expansion driving the Universe's infancy allows regions which were large enough to have been spanned by light signals just before inflation commenced, to have grown larger than the entire visible part of the Universe today (Figure 8.7). In the absence of this period of inflationary expansion, those coordinated regions would have grown only to no more than a metre in size today – falling short of an explanation of the extent of the uniformity of the astronomical universe by 10^{24} metres.

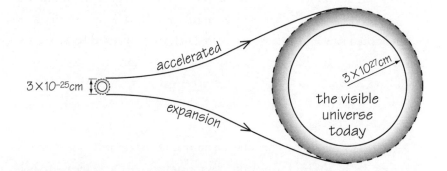

Figure 8.7 *Inflation grows a region bigger than the visible part of the Universe today from a region small enough to be coordinated by light signals near the beginning of the expansion. This offers an explanation for the uniformity of the visible Universe today.*

Remember that the key idea behind Einstein's general theory of relativity was that the presence of mass and energy in space will cause it to be curved. This curvature we imagined to be like the undulations caused by heavy objects on a rubber sheet. If the universe is very irregular before inflation begins it is as if the rubber sheet of the universe is very lumpy and bumpy. When inflation begins it creates a stretching effect, driven by the accelerating expansion, which will iron out all the hills and valleys. It will also make the whole sheet look locally rather flat. If you draw a small square on the surface of a balloon as it is inflated then the square will appear to get flatter and flatter as the balloon is inflated. The universe with the critical rate of expansion is one whose space is flat and uncurved at any time. The other universes that expand faster and slower have negatively and positively curved spaces undergoing expansion, respectively. In both cases they will locally look more and more like a flat surface the more inflationary expansion they have experienced. Almost all[8] curved surfaces look locally like flat ones when surveyed over small distances.

Inflation kills many birds with one stone. It explains why it is natural for the Universe to be expanding on a trajectory very close to the critical divide today; it explains why the Universe is on average so smooth from place to place and from one direction to another when we survey its density, temperature and expansion rate. Inflation enables the Universe to maintain life-supporting conditions for the billions of years needed for stars to form and biochemical processes to produce replicating molecules and complex organisms. If the expansion had not tracked the critical divide so closely then it would either have peeled off and collapsed back to a big crunch of uninhabitably high density long before stars could form, or it would have expanded so rapidly that neither galaxies nor stars could have condensed out to create the building blocks and stable environments needed for life (Figure 8.8).

Thus, the complexity of the vacuum that makes inflation possible lies at the root of the uniformity of Nature and allows the Universe to persist for billions of years, displaying conditions that are conducive to the formation of stars and biochemical elements.

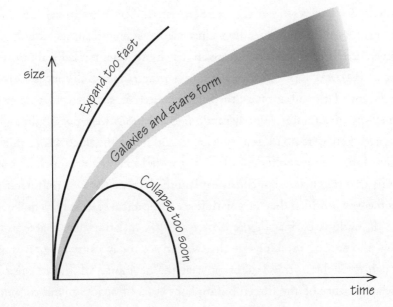

Figure 8.8 *Universes that expand too slowly will collapse back to a big crunch before galaxies can form; universes that expand too quickly do not allow islands of matter to condense out into galaxies and form stars.*

VACUUM FLUCTUATIONS MADE ME

"The universe is merely a fleeting idea in God's mind – a pretty uncomfortable thought, particularly if you've just made a down payment on a house."

Woody Allen[9]

If the game of musical vacuums that leads to inflation had resulted in a universe that was *perfectly* smooth and featureless then things would have turned out pretty dull. There would be little to write home about; indeed no one *to* write home. Although our Universe is extremely close to uniformity, it is not perfectly so. There are small deviations from uniformity in the density of matter in space in the form of stars and galaxies and great clusters of galaxies – even clusters of clusters.[10] In order to explain their presence, we need the expanding Universe to emerge from its early high-temperature history with variations in density that are typically about one part in 100,000 above the average over a wide range of distances. Before the advent of the inflationary theory the source of such irregularities was something of a mystery. Purely random fluctuations were not of the right size, and there were no ideas as to what the origin of the fluctuations might be, let alone their magnitude. Inflation provided a new and compelling possibility that might simultaneously explain the level of the non-uniformities and the way in which they vary with the astronomical scale surveyed.

If we look back at Figure 8.7 we see how inflation may enable us to 'grow' the part of the Universe that we can see today from a region small enough for light to travel across it near the beginning of the expansion. The appearance of the fifteen billion light years of space around us today derives from a tiny region. We are its greatly expanded image. If smoothing processes were perfect and that tiny region started off perfectly smooth, then its subsequent inflation would create a large and perfectly smooth region. But, alas, *perfectly* smooth means no little islands of matter that expand more slowly than the rest, and which break away from the universal expansion to form galaxies and stars which initiate nuclear reactions and the supernovae from which come the biological elements like carbon. All would be cosmic sameness. No structure, no stars, just perfect undisturbed symmetry.

Fortunately for us, this cannot quite be. There must exist fluctuations of quantum uncertainty in the vacuum. The scalar fields whose slow changes can drive the acceleration of the Universe must have zero-point motions. Just as Heisenberg's Uncertainty Principle forbids us from ever saying that a box is empty, so it forbids us from ever saying that the density or the temperature of the vacuum is perfectly smooth. There must always exist some quantum vacuum fluctuations. So when inflation occurs it will also act upon the very small deviations from perfect uniformity that the zero-point fluctuations create. They will be stretched by the inflationary expansion and left, like scars, on the face of the Universe, tracing small variations in its density and temperature out to the largest astronomical distances. Remarkably, we can predict the form that these fluctuations must take and their fate during the inflation process. These vacuum fluctuations will eventually lead to the aggregation of matter into galaxies and stars, around which planets can form and life can evolve. Without the vacuum the book of life would have only blank pages.

There are two things we need to predict about these stretched vacuum fluctuations: how intense they will be on average and how their undulations should vary with the distance surveyed. Unfortunately, the first of these questions does not yield a clear-cut answer that we can go out and test. Inflation is an appealing idea because the more you look into what will happen to elementary particles during the first moments of the Universe's expansion the harder it is to *avoid* inflation. Almost any hypothetical scalar field will do the trick. Inflation is a rather robust consequence that does not depend on very special conditions. However, the intensity of the fluctuations that are dredged up from the vacuum and expanded depends on knowing the mass of the particular scalar matter field that did the inflating. All we can do is reverse-engineer the situation to calculate what intensity level would be needed to grow the galaxies that we see, and determine the mass of scalar field that gets it right. This requires a little work because galaxies do not appear from the fluctuations ready-made. The fluctuations can begin with a very low intensity, but gradually they will become more pronounced. Regions which contain a little more matter than average will attract still more material towards them at the expense of the others – a sort of gravitational Matthew Effect[11] that 'unto he who has shall more be given', which astronomers call gravitational instability. The process will snowball and eventually produce dense islands of matter in an almost smooth background universe.

Working backwards we can calculate how small the initial non-uniformities need to be if they are to grow into the observed stars and galaxies in the time available since the Universe became cool enough for atoms to form.[12] This tells us that the vacuum fluctuations need to be approximately a few parts in 10^5 in intensity. We have a double check on this from the satellite observations of the microwave background radiation from the Big Bang. The ancient vacuum fluctuations will have left scars in this radiation long before the galaxies ever formed. Astronomers have been searching for these tell-tale imprints from the past ever since the radiation was first discovered in 1965. They have finally been found by NASA's Cosmic Background Explorer (COBE) satellite orbiting high above the distorting influence of the Earth's atmosphere. What it sees confirms that fluctuations of the required level were indeed present at the stage when the heat radiation from the Big Bang began its journey towards us. This tiny measured fluctuation level of a few parts in 10^5, mapped over parts of the sky separated by more than about ten degrees, now acts as a guide to physicists as they try to winnow down the possible scalar matter fields that could have been responsible for inflating the vacuum long ago.

Fortunately, that is not all that can be said. Although we cannot predict the level of the fluctuations expected to emerge from inflation, because it is so sensitive to the identity of the field driving the inflation, we can predict the way in which the pattern of fluctuations should vary with the astronomical scale surveyed. This turns out to be far less sensitive to the identity and properties of the inflating field. There is a simple and most natural case in which the fluctuations have a democratic form, contributing the same curvature of space on every dimension over the very largest astronomical scales. By comparing parts of the sky separated by more than about ten degrees (the face of the full Moon spans about half a degree), the COBE satellite has confirmed these expectations to high accuracy. This is encouraging, but the greatest interest is reserved for much smaller scales which encompass the fluctuations from which the observed clusters and galaxies will have formed. Very recently, these have been extensively mapped for the first time. The results of Boomerang, a balloon experiment launched from the South Pole, show a very close match with the predictions for expanding universes that are very close to the critical divide. In Figure 8.9, the Boomerang results are shown against a continuous curve which is a theoretical prediction of the form expected in a universe that is just slightly denser than the critical value. The key feature that the observers were looking for is the peak in the amplitude of the temperature fluctuation close to separations on the sky of one degree. Its precise loca-

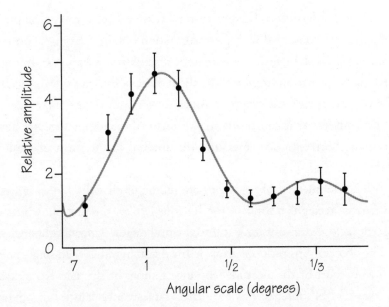

Figure 8.9 *The variation of temperature fluctuations in the microwave background radiation found by the Boomerang project.*[13] *A fit to the data by an almost critical expanding universe's predictions for these fluctuations is shown. The angular location of the first peak in the fluctuation is our most sensitive probe of the total density of the Universe.*

tion is the most accurate probe of the total density of the Universe. This is the first time that this peak has been unambiguously observed. There is a suggestion that there is a second, lower peak in the data at smaller angles, but more accurate observations will be needed to make a convincing case for its presence.

In 2001 a further satellite probe, MAP (the *Microwave Background Explorer*), will be launched by NASA to pin down the shape of the fluctuation curve with far greater precision over a wider range of sky angles. In 2007, an even more powerful detector, *Planck Surveyor*, will be launched by the European Space Agency to scrutinise these variations in exquisite detail. The potential pay-off from these two missions is huge. They will enable us to determine whether the distinctive relics of inflation do indeed exist in the Universe and probe directly the vacuum fluctuations emerging from the Big Bang.

These observations become even more powerful cosmological probes when they are combined with the information obtained from the observations of very distant supernovae that we discussed in Chapter 6. In Figure 8.10, the information from both of these observations are shown together. The vertical axis of the graph measures the amount of the energy density in the Universe that can reside in the form of quantum vacuum energy whilst the horizontal axis measures the amount in the form of ordinary matter.

The Boomerang observations are telling us that the Universe lies in the narrow triangular band in the bottom left of the picture, whilst the supernovae observations force it into the oval region lying at right angles to it. The observations pick out areas of the diagram rather than single points or lines because of the measurement uncertainties of the data. Remarkably, the two sets of observations have their largest uncertainties in opposite directions, so in combination they can pin down the Universe by their overlap with far greater uncertainty than when taken singly. We see that the overlap region requires that the vacuum energy contribution to the Universe is very significant. It cannot be anywhere near zero if these observations are both correct.

INFLATION ALL OVER THE PLACE

"I never predict anything and I never will do."

Paul 'Gazza' Gascoigne[14]

Soon after the benefits of a bout of cosmic inflation were first recognised, it became clear that the consequences were vaster than had been imagined. Suppose that, just before inflation occurred, the Universe was in a pretty chaotic state. It may have contained a huge number of scalar matter fields, all different, some of them possibly affecting one another in complicated ways. Each could have a different potential landscape down which it would fall, starting out at different speeds and slowing at different rates. This anarchic scenario of 'chaotic' inflation creates for us a picture of a universe

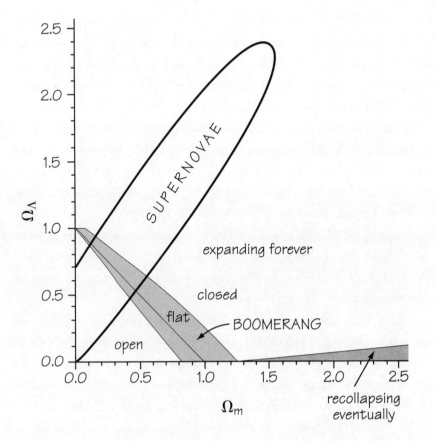

Figure 8.10 *The limits on the relative contributions to the total energy density in the observable Universe contributed by matter (Ω_m) and by the vacuum (Ω_Λ), the latter in the form of a lambda stress.[15] The 'Supernovae' region is compatible with observations of the recession of distant supernovae taking part in the expansion of the Universe. The 'Boomerang' region is consistent with the Boomerang balloon flight observations of the small-scale fluctuations in the microwave background radiation. The 'flat' line separates open universes from closed universes. Also marked is the region which allows the Universe to collapse back to a 'big crunch'. This latter region is incompatible with both data sets. The overlap region compatible with Supernovae and Boomerang requires a significant, non-zero contribution by the vacuum energy to the total density of the Universe.*

in which every region that is small enough to be smoothed by light signals could have undergone a period of inflation. The amount of inflation that each region will undergo will be random: some regions will experience a lot of inflation and ultimately expand to become very large, whilst others will barely inflate at all and their expansion could be reversed into contraction very soon afterwards. It is like a foam of bubbles being randomly heated so that some of the bubbles expand a lot, others a little. The most short-lived inflationary histories create regions which don't expand long enough to see stars form and produce the building blocks of life. These still-born 'bubbles' will contain no astronomers. Some of the large, long-lived bubbles may expand for billions of years, creating room and time for stars to form the building blocks of biochemical complexity. It is only in one of these big, old bubbles that observers like ourselves can be around to take stock of the cosmic scene.

Seen in this light, inflation has an air of inevitability about it. If the Universe is infinite in extent then anything that has any chance of occurring will be occurring somewhere, and so somewhere there will be a region where there is a matter field whose potential-energy landscape is shallow enough for a very slow change to create a lot of accelerated expansion. Even if this is an unlikely situation (although there is no reason to think that it is), it will still happen in some places and we will find ourselves residing in one of them.

This scenario makes our picture of the geography of the Universe vastly more complex. Ever since Copernicus, we have been educated to assume that our location in the Universe is not special. Our observations of the visible Universe show it to be extremely similar from place to place and from one direction to another on average. Copernicus implies that we should see the same level of uniformity on average from any cosmic vantage point. Thus we should expect the Universe to be roughly similar everywhere. There were always sceptics who did not trust this argument and pointed out that we could never be sure that things are not very different in the Universe beyond our visible horizon, fifteen billion light years away. Despite their logical correctness, these commentators had no positive reason for believing that the far-away Universe was different. The chaotic inflationary Universe is revolutionary because for the first time it provides us with a positive reason to expect the Universe to be very different in structure beyond our visible horizon. Even if the Universe did not begin chaotically and there is only one scalar energy field available, the random variations in its behaviour from place to place are enough to create many different inflated regions. At present, we must assume that we can just see the smooth, nearly flat, interior of part of one of them. If we waited long enough, maybe trillions of years in the future, the expansion might reveal the first glimmerings of a region with a quite different structure swimming slowly into view. The little variations in the structure of the vacuum from place to place will have been amplified from microscopic scales to the vastness of extragalactic space. The universality and diversity of the vacuum landscape in the Universe has the scope to expand to become the direct source of the entire cosmic array of light and darkness, space and matter, planets and people. It makes the Universe more complicated than we imagined.

MULTIPLE VACUUMS

"It does not do to leave a dragon out of your calculations, if you live near him."

J.R.R. Tolkien

We have seen how the valleys of the potential energy landscape can have many different minima. They may all have the same levels or they may be different. The possibility of different vacuum states is far-reaching because if our Universe possesses different possible vacuums it means that the constants of physics, quantities which measure the strengths and properties of the forces of Nature, need not be uniquely determined. They could have fallen out differently, and may even have done so, in some of those distant domains where different amounts of inflation occurred. If the vacuum energy landscape for the Universe has a single minimum then the basic constants of physics and the form of the laws governing the forces of Nature must be the same everywhere.

Let's look at the situation with many vacuums more closely. Suppose that the early Universe is inhabited by a matter field that moves in a potential energy landscape that is corrugated, with many minima, as in Figure 8.11. Imagine that the cooling down of the Universe, soon after the expansion begins, scatters the field to some random point in this sinuous landscape. It will then start to roll down the slope on which it finds itself towards the local vacuum state. In other parts of the Universe the field will find itself in different valleys and it will end up rolling (perhaps slowly) into a different vacuum state. The consequences of such diversity would be very far-reaching. Each of these vacuums will correspond to a future world with different forces of Nature. One region might inflate into a state in which gravity is the strongest force of Nature that exists. There would be no stars, no nuclear reactions, no chemistry and no life. There is a deep and direct connection between the multiplicity of vacuums and the uniformity in the Universe of those features of its legislation that we have come to call the constants and laws of Nature. This is not the end of it. Even the number of dimensions of space that inflate and become astronomically large can differ from valley to valley along with the constants and forces of Nature. In recent years, physicists have begun to take seriously the possibility that space (and even time) might contain more dimensions than we habitually experience. Somehow physics looks simpler and naturally unified at high temperatures in worlds which possess more than three dimensions. In order to reconcile such a higher dimensional universe with the space that we observe, it is necessary to assume that all but three of the

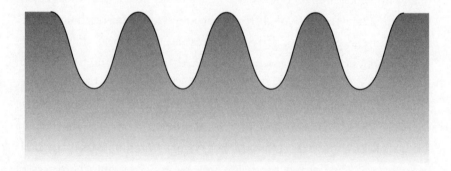

Figure 8.11 *A sinuous vacuum landscape with many minima.*

dimensions are imperceptibly small. No one knows how this happens. Perhaps inflation can be selective in some as yet unknown way, allowing only three dimensions of space to inflate and become astronomically large, whilst the others stay imperceptibly small. If a process like this does operate, it might only work when three dimensions become large; or perhaps it is entirely random, so that the number of large dimensions varies all over an infinite universe. Again, we have good reason to believe that living observers will most likely find themselves in a region possessing three large dimensions of space and a single arrow of time. Some of the consequences of different dimensions of space and time are shown in Figure 8.12.

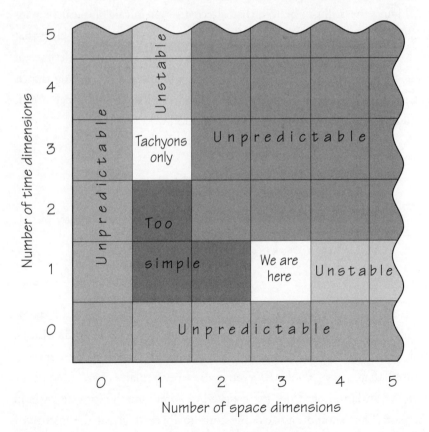

Figure 8.12 *Universes with different numbers of dimensions of space and time have unusual properties that do not look conducive to complex information-processing and life except when there is one dimension of time and three large dimensions of space.*[16]

Such possibilities change our entire conception of our place in the Universe. We know that our existence is only possible because of a number of fortuitous apparent coincidences between the values of different constants of Nature. If the values of those constants are unchangeably programmed into the formation of the Universe then we might have to conclude that it was rather good luck that they have fallen out to permit life as they have – of course, if they had not done so we would not be here to argue about it.

Alternatively, we might try to argue that life is possible in a multitude of ways other than by means of DNA molecules based upon the properties of elements like carbon, nitrogen and oxygen. Actually, many scientists (including the author) believe that alternative chemical, physical or nanotechnological bases for complexity are very likely, but it is not clear that they can evolve life spontaneously[17] on a timescale less than the lifetimes of stars. One day we may develop a form of information processing that is sufficiently complex to merit the name 'life' or 'artificial intelligence', but it would not have arisen by natural selection alone.

ETERNAL INFLATION

"We know what we are, but know not what we may be."

William Shakespeare

Soon after it was realised that a 'chaotic' vacuum landscape could give rise to different degrees of inflation all over an infinite universe, Andrei Linde and Alex Vilenkin, both Russian physicists now working in America, realised that things could be even more spectacular. These ubiquitous bouts of inflation need not be relegated to some time billions of years in the past. They should be occurring continually throughout the history of the Universe. Even today, most of the Universe beyond our visible horizon is expected to be in a state of accelerating inflation.

Although it appears that our hypothetical scalar field will just roll down the slope of the potential landscape towards the nearest vacuum, the quantum picture of the vacuum introduces tiny fluctuations which make the field zigzag up and down as it moves down the hill. Remarkably, it is very likely that the zigzagging will predominate over the simple downhill rolling and occasionally make the field move up the valley instead of down. It is like a very slowly flowing river moving down a very shallow gradient. In addition to this steady flow there will be a random to-and-fro motion of flotsam on the water surface. If the overall flow is slow enough and the wind strong enough, some of the debris can occasionally move upstream. In the cosmological case this tendency leads to the production of further inflation within sub-domains of the Universe which have already undergone inflation (see Figure 8.13).

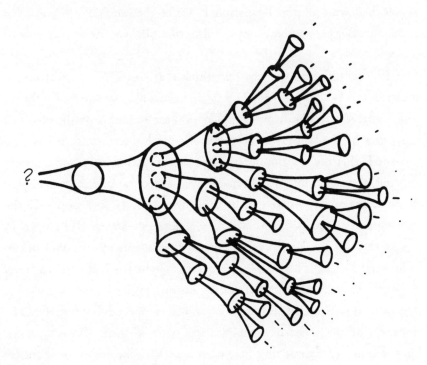

Figure 8.13 *Eternal inflation.*

The spectacular effect of this is to make inflation self-reproducing. Every inflating region gives rise to other sub-regions which inflate and then in turn do the same. The process appears unstoppable – eternal. No reason has been found why it should ever end. Nor is it known if it needs to have a beginning. As with the process of chaotic inflation, every bout of inflation can produce a large region with very different properties. Some regions may inflate a lot, some only a little; some may have many large dimensions of space, some only three; some may contain the four forces of Nature that we see, others may have fewer. The overall effect is to provide a physical mechanism by which to realise all, or at least almost all, possibilities some-where in a single universe.

This is a striking scenario. It revolutionises our expectations about the complexity of the evolution, past and future, of the Universe in the same way that the possibility of chaotic inflation did for our picture of its geography. There have often been science-fiction stories about all possible worlds displaying all possible permutations of the values of the constants of Nature. But here we have a mechanism that can generate the panoply of choices.

Eternal inflation was not something that cosmologists went out to construct deliberately. It turned up as an inevitable by-product of a theory which offered a straightforward explanation for a number of the observed properties of the Universe. Future astronomical observations will be able to test whether the structure of the radiation fluctuations in the Universe are consistent with inflation having played a decisive role in determining the structure of our visible part of the Universe. So far, unfortunately, the entire grand scheme of eternal inflation does not appear to be open to observational tests. We cannot see further than a distance of about fifteen billion light years. This is the distance that light has had time to travel since the apparent beginning of the expansion that we are now witnessing. The other different domains of inflation will be beyond that horizon. The finiteness of the speed of light insulates us from them. One day, when huge amounts of cosmic time have passed, perhaps the observers of the far future will be privileged to witness the first appearance of one of these strange islands of the Universe, where inflation is still going on or where the laws of physics are very different. Overall, the Universe is likely to be in a steady state, but populated by many little inflating bubbles, each spawning a never-ending sequence of 'baby universes'. Most of the Universe will be undergoing inflation at the moment. We live in one of the regions where inflation stopped in the past and we could not exist if it were otherwise. An inflating region expands too fast for galaxies and stars to form. Those essential steps in the path towards setting up life-supporting environments must wait until inflation has ended. However, if the Supernova observations are correct we may be witnessing the recent resumption of inflation in our part of the Universe. If so, we don't know why this is happening.

This revolution in our conception of the Universe sees us as inhabitants of a large domain that has arisen in a cosmic history with neither beginning nor end, where the special requirements for stars and chemistry and life to evolve are met. This local part of the Universe that has inflated to contain our visible portion of the Universe is just part of the story. Elsewhere, the Universe is predicted to be very different. Globally, our conception of the Universe has been transformed and we must expect that what we can see is not likely to be representative of the whole. All the complexity that we expect to define the totality of the Universe around us is a reflection of the structure of the vacuum. It is a bottomless sea of energy for expanding universes to produce offspring in the form of sub-regions that go their own way, becoming larger and cooler, ultimately creating within themselves the conditions for further baby universes to be born.

At first, these events of inflationary reproduction appear to be spawning something out of nothing. In fact, the situation does nothing of the sort. We might think that if a whole sub-region of universe appears and starts to expand then we must be violating one of the great conservation laws of physics. The most familiar is the conservation of energy. It was discovered in the last century that in all natural processes, the quantity that we call 'energy' is conserved. We can change its form, shuffle it around in different ways, use it to turn mass into radiation and vice versa, but when all is said and done, after we do the final accounting we should always find that the total energy comes out the same. So we might think that if we go from 'no universe' to 'universe' we are getting something — energy — for nothing and our fundamental conservation law is being broken. However, things are not so simple. Energy comes in two forms. Energy of motion is positive but potential forms of energy are negative. The latter is possessed by any body that feels an attractive force, like gravity.

Universes and inflating domains within universes have very surprising properties when we start to inquire about their energies. Einstein's theory of general relativity ensures that the total of the positive values of the energies of all the masses and motion within them is *exactly* counterbalanced by the sum of the negative potential energies contributed by the gravitational forces between them. The total energy is zero. An expanding region can appear without any violation of the conservation of energy. This is a rather striking conclusion. It shows how a large amount of inflationary expansion can be underwritten by drawing on a large reservoir of negative potential energy.[18]

INFLATION AND NEW LAMBDA

"I have yet to see any problem, however complicated, which when you looked at it in the right way, did not become still more complicated."

Poul Anderson

In Chapter Six we first encountered the deep mystery of the lambda problem. Einstein had found that the force of gravity that Newton uncovered should be partnered by another piece that increases over large distances. Despite later regretting ever letting this genie out of the bag, saying that it was 'the biggest blunder of my life', and urging scientists to ignore it, Einstein's arguments against his creation were never persuasive. In 1947, he wrote despairingly in a letter to his fellow pioneering cosmologist, Georges Lemaître, that

"Since I have introduced this term I have always had a bad conscience. But at that time I could see no other possibility to deal with the fact of the existence of a finite mean density of matter. I found it very ugly indeed that the field force law of gravitation should be composed of two logically independent terms which are connected by addition. About the justification of such feelings concerning logical simplicity it is difficult to argue. I cannot help to feel it strongly that I am unable to believe that such an ugly thing should be realised in Nature."

You might not like it. You might wish that it would simply go away. But unfortunately, as yet, there seems to be no good reason to exclude it.

Until quite recently, most physicists who worried about this problem were looking for a missing insight to prove that lambda must be zero. They were persuaded of the rightness of this approach by the unnatural situation that is created by the existence of a force like this that 'just happens' to become noticeable in the Universe around the epoch when we are living in the Universe, about fourteen billion years after the expansion began. But we have just witnessed a change of attitude. Astronomers have found strong evidence for the existence of a non-zero lambda force. Its size means that it has come to govern the rate at which the Universe is expanding at about the time when galaxies were still forming – what astronomers would call 'quite recently'. From the theoretician's point of view this is very odd. Lambda not only exists, but has a special value that makes it come into play near the epoch when life develops in the Universe. The only consolation is that, if these observations are correct, there is now a very special value of lambda to try to explain. The right explanation has a very particular target to shoot at. One can imagine a lot of spurious arguments that manage to 'explain' why lambda is zero but not so many that can come up with the unusual observed value.

Inflation has solved a lot of our other puzzles; can it help us with lambda? Unfortunately, it is hard to see how inflation can help. We have already seen how the lambda stress is like a vacuum energy in the Universe.

If we look at our potential energy landscape for the scalar field that is driving the inflationary expansion we can relate the presence of lambda to a special property of its topography. In the examples that we have drawn (like Figure 8.5, for example), the level of the minimum that defines the true vacuum state has been placed at a zero value. But there was no reason to do that. It was just artistic licence. The final minimum energy value could have been placed at any level above the zero line. Our knowledge of physics does not tell us where it should be. However, if this level is above the zero line, as in Figure 8.14, then it will leave an energy in the Universe that behaves exactly like the lambda stress. Its height above the zero line will determine the magnitude of the lambda force.

When one looks at the numbers, the situation becomes even more perplexing. The effect of lambda grows steadily with respect to the familiar Newtonian force of gravity as the Universe gets bigger. If it is only recently becoming the dominant force, after billions of years of expansion of the Universe, it must have started out enormously smaller than the Newtonian force. The distance of that final minimum energy level in Figure 8.14 from the zero line in order to explain the value of lambda inferred from the supernova observations is bizarre: roughly 10^{-120} – that is, 1 divided by 10 followed by 119 zeros! This is the smallest number ever encountered in sci-

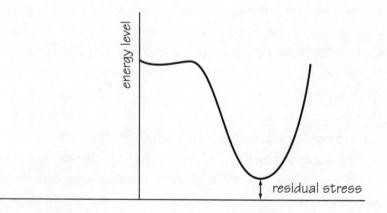

Figure 8.14 *The height of the minimum above the zero line determines the residual value of the lambda stress in the Universe.*

ence. Why is it not zero? How can the minimum level be tuned so precisely? If it were 10 followed by just 117 zeros, then the galaxies could not form. Extraordinary fine tuning is needed to explain such extreme numbers. And, if this were not bad enough, the vacuum seems to have its own defence mechanism to prevent us finding easy answers to this problem. Even if inflation does have some magical property which we have so far missed that would set the vacuum energy *exactly* to zero when inflation ends, it would not stay like that. As the Universe keeps on expanding and cooling it passes through several temperatures at which the breaking of a symmetry occurs in a potential landscape, rather like that which occurs in the example of the magnet that we saw at the beginning of the chapter. Every time this happens, a new contribution to the vacuum energy is liberated and contributes to a new lambda term that is always vastly bigger than our observation allows. And, by 'vastly bigger' here, we don't just mean that it is a few times bigger than the value inferred from observations, so that in the future some small correction to the calculations, or change in the trend of the observations, might make theory and observation fit hand in glove. We are talking about an overestimate by a factor of about 10 followed by 120 zeros! You can't get much more wrong than that.

All our puzzles about whether or not lambda exists and, if so, what is responsible for giving it such a strange value, are like questions about the inflationary scalar field's potential landscape. Why is its final vacuum state so fantastically close to the zero line? How does it 'know' where to end up when the scalar field starts rolling downhill in its landscape? Nobody knows the answers to these questions. They are the greatest unsolved problems in gravitation physics and astronomy. The nature of their answers could take many forms. There could exist some deep new principle that links together all the different forces of Nature in a way that dictates the vacuum levels of all the fields of energy that feel their effects. This principle would be unlike any that we know because it would need to control all the possible contributions to lambda that arise at symmetry breakings during the expansion of the Universe.[19] It would need to control physics over a vast range of energies.

Alternatively, there could be a less principled solution in which the lambda stress is determined completely randomly. Although huge values of lambda are the most probable and persistent, they give rise to a universe that expands too fast too early for stars and galaxies and astronomers ever to appear. If we were casting our eye across all possible universes displaying all possible values of the lambda stress, it could be that those, like our own, with outlandishly small values are self-selected from all the possibilities by the fact that they are the only ones that permit observers to evolve. In fact, if lambda were just one hundred to a thousand times bigger than the observations claim, the sequence of events that led to us might well be prevented. Bigger still and they definitely would be. This type of approach, while it may be true, can never predict or explain the exact value of lambda that we have observed, because life is not so sensitive to the value of lambda that, say, doubling its value would make life impossible.

FALLING DOWNSTAIRS

". . . but we shall all be changed, in a moment, in the twinkling of an eye . . ."

St Paul[20]

The picture of many vacuums that may characterise the forces and interactions of Nature gives rise to the possibility of inflation. There are many options as to how the change from one ephemerally stable vacuum to another true vacuum might occur and we have no knowledge as yet of the identity of the scalar field which might be the culprit.[21] In this way of looking at vacuum, we have so far imagined that the vacuum state in which we now find ourselves is a deep and stable one, a 'true' vacuum. The lowest of the low.

What if we are not in such a vacuum basement? It is entirely possible that the state of the Universe in which we find ourselves is that of a temporarily stable, or 'false', vacuum. Instead of being on the ground floor of the vacuum landscape, we may be higher up, in a state that is only stable for a period of time. That period is pretty long, because the Universe seems to have possessed the same general laws and properties for about fourteen billion years. But one day things may change very suddenly, without the slightest warning. The situation could be like that pictured in Figure 8.15. If inflation left us lodged in on the shallow ledge in the potential landscape shown in Figure 8.15, then we might suddenly find ourselves nudged over the brink and on the way down to a lower minimum. That nudge might be supplied by very high energy events in the Universe. If collisions between stars or black holes generated cosmic rays of sufficiently high energies, they might be able to initiate the transition to the new vacuum in a region of space.[22] The properties of the new vacuum will determine what happens next. We could find ourselves suddenly falling into a vacuum state in which all particles have zero mass and behave like radiation. We would disappear in a flash of light without warning.[23] The way in which our form of bio-chemical life relies on rather particular coincidences between the strengths and properties of the different forces of Nature means that any change of vacuum state would very likely be catastrophic for us. It would leave us in a new world where other forms of life might be possible but there is no rea-

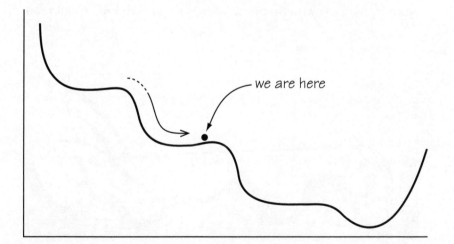

Figure 8.15 *A potential energy landscape with many shallow minima may gradually evolve downstairs from one minimum to another over billions of years. We may not yet have reached the bottom.*

son why they should be a small evolutionary step away from our own bio-chemical forms.

This picture of the vacuum landscape is a speculative one. We do not know the overall form of the landscape well enough to be able to tell whether we are already on the ground floor or whether there are other vacuums downstairs into which the states of matter in our locale can fall, either accidentally or deliberately. As one contemplates this radical possibility of an unannounced change in some of the basic properties of the forces of Nature, it is tempting to portray it as the ultimate extension of the idea of punctuated equilibrium that Niles Eldridge and Stephen Jay Gould[24] have promoted. They proposed the course of biological evolution by natural selection on Earth proceeds by a succession of slow changes interspersed by sudden jumps rather than as a steady ongoing process. Indeed, we can characterise it as a movement through a landscape with many hills and valleys in which a force is dragging someone along. The pattern of change under these circumstances is for a slow climb up each hill but when the top is reached there will be a sudden jump across to the side of the next hill and another spell of steady hill-climbing (see Figure 8.16). If the Universe follows this lead there may be a shock for our descendants in aeons to come. As with the puzzle of why the lambda force should come

Figure 8.16 *A typical evolution in a landscape with many minima when there is a force acting. The dog climbs the slopes slowly to the top and then suddenly jumps across to a point on the next ascent and begins slowly climbing uphill again.*[25]

into play so close to our time, so we might regard it as unlikely that the epoch at which the fall 'downstairs' could occur should be close to the time of human existence in the Universe – unless, of course, there is a link with lambda, or the presence of life can do something inadvertent to precipitate the great fall downstairs. Prophets of doom: do not give up hope.

BITS OF VACUUM

"Cats, no less liquid than their shadows,
Offer no angles to the wind.
They slip, diminished, neat, through loopholes
Less than themselves."

A.S.J. Tessimond[26]

At the start of this chapter, we described a vacuum landscape which was three-dimensional. Imagine a Mexican hat with a shallow valley at the top of the hat and an entire circle of minima all at the same level at the bottom of the hat's brim, as in Figure 8.2. It is possible to move around the circle of vacuum states in the trough at the bottom of the hat without changing energy. In 1972, the British physicist Tom Kibble[27] realised that the possible existence of vacuums with continuous interrelationships of this sort meant that changes in their shape could occur as the Universe cooled, which would create structures in the Universe which retained memory of the energy of the Universe at the time when they formed. They are pieces of vacuum. Depending on the shape and pattern of the possible multiple vacuums they could have three simple forms. There can be lines of vacuum energy, either closed loops or never-ending lines, called '*cosmic strings*'.[28] There can be sheets of vacuum energy which extend for ever, called '*walls*', and there can be finite-sized spherical knots of vacuum energy called *monopoles*. The strings have a thickness given by the quantum wavelength corresponding to the energy of the Universe when the symmetry breaking that created them took place. Similarly, walls are sheets of vacuum energy with a thickness determined by this quantum wavelength.

These three vacuum structures have proved to be perennially fascinating to astronomers ever since their possible existence was first recognised. It was soon realised that if they could exist then their impacts on the Universe are very different. Walls were only an optional structure in the theories of matter at very high energies that were being explored. This was fortunate because walls are a disaster for the Universe. A single vacuum wall stretched across the visible Universe would exert a devastating gravitational force on the expansion of the Universe and produce huge differences in the intensity of radiation from different directions in the Universe. Evidently, from our observations of the smoothness of the radiation and the expansion, we can conclude that we are not in the presence of cosmic domain walls. This deduction is an example of how an astronomical observation can provide a constraint on the possible properties of the unified theory of the forces of Nature at very high energies which are beyond the reach of the energies attainable by direct experiments.

The next candidate to be considered is the monopole. These are far more problematic. Unlike walls, monopoles appeared to be inevitable in any reasonable theory of how the Universe changed from the high-temperature environment of the Big Bang to the present low-temperature world that we inhabit. If the forces of electricity and magnetism are to exist in our world today then monopoles must be formed in the early Universe. Alas, their presence is another potential disaster. A monopole should form inside every region that light signals have had time to cross from the beginning of the expansion of the universe to the time when the monopoles can appear. Such regions are very small because monopoles are very massive by the standards of elementary particles, and appear in pairs when the universe is very energetic and very young. This means that the region of the universe that eventually expands to become the fifteen-billion light-year expanse that we call the visible Universe today will contain a huge number of these monopoles. When we add up the masses of all the monopoles that we should find, their total mass turns out to be billions of times greater than that of all the stars and galaxies put together. This is not the Universe that we live in. Indeed, it is not a universe that we *could* live in.

In the mid-1970s, this 'monopole problem' was a serious dilemma for physicists trying to develop a unified theory of the different forces of Nature. The candidate theories had many attractive features that offered explanations for particular properties of the Universe, most notably why it displayed such an overwhelming excess of matter over antimatter. But they all predicted a monopole catastrophe. Experimental physicists, on the other hand, didn't see these monopoles. What happened to them?

It was this problem that first led Alan Guth, then at Stanford University, to the theory of the inflationary Universe. He saw that initiating a period of accelerated expansion would solve the monopole problem in the same way that it solved the problem of the smoothness of the Universe. The inflationary surge of acceleration enabled the whole of our visible Universe to expand from a region that was once small enough for light signals to keep it smooth and coordinated except for the small zero-point fluctuations. A monopole forms every time a mismatch occurs in the direction in which vacuum energy fields are pointing when the universe cools to the energy level of the monopoles. Mismatches produce 'knots' in the vacuum energy that manifest themselves as monopoles. These knots can only be ironed out over regions that are small enough for light signals to traverse in the time before the appearance of the monopoles. Guth saw that inflation would enable the whole of our visible Universe today to be encompassed by a region that was once small enough to contain perhaps only one knot of vacuum energy and a single monopole. Their effect on the expansion of our visible Universe would then be utterly negligible and we have a natural explanation for the mysterious cosmic scarcity of monopoles.

What Guth was proposing was that the monopoles are not prevented from forming (as many others were trying to find ways of demonstrating at the time), nor were they annihilated in some way after they formed (as others had also tried to show): they are just moved so far by the expansion that they are beyond the horizon of our visible Universe today. Just as the smoothness of our visible Universe is a reflection of the smoothness of the small domain from which it inflated so its lack of monopoles derives from the smooth, unknotted character of the vacuum fields within the same domain.

Historically, the prime motivation for devising the theory of inflation was the resolution of the monopole problem. An added initial bonus was to provide an explanation for the smoothness and flatness of the visible Universe. However, as time has gone on, the focus of interest has switched to the prediction of inflation that the zero-point fluctuations will be inflated to produce little irregularities from which galaxies can form, for it is here that a critical observational test of the theory will soon be made.

This leaves one more vacuum structure for us to evaluate: the strings. Cosmic strings turn out to be far more interesting than walls or monopoles. Whereas walls and monopoles both threaten to overpopulate the Universe with unwanted mass, and have to be eradicated early on, cosmic strings are more subtle. They will start by threading the Universe with a great network of lines of vacuum energy, like a web of cosmic spaghetti. As the expansion of the Universe proceeds, the network behaves in a complicated fashion. Whenever intersections of string occur, the string reorganises itself by exchanging partners, as shown in Figure 8.17.

The trend is for the network to produce lots of little loops of string at the expense of long lines of string that run across the Universe. Once a small loop is formed it is doomed to dissolve. It will oscillate and wriggle, gradually radiating all its energy away in the form of gravitational waves. If we think of Einstein's picture of curved space, then the wiggling of the loops of string creates ripples in the geometry, which spread out at the speed of light, taking away the string's energy like waves on a pond surface. In Figure 8.18 a computer simulation of an expanding box of cosmic strings is shown.

The behaviour of the string network over the history of the Universe is tantalising. It appears that the presence of the loops and lines of string

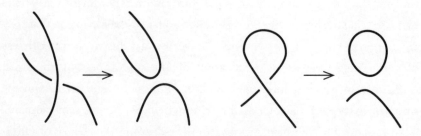

Figure 8.17 *Cosmic strings exchange loops when they intersect.*

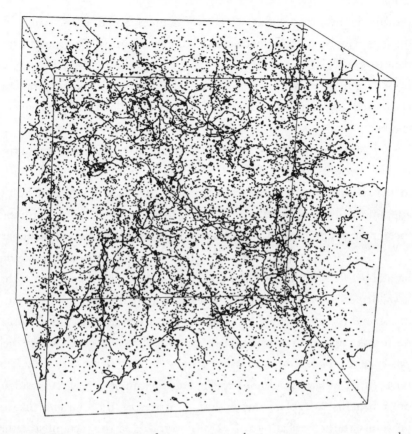

Figure 8.18 *A computer simulation of a network of cosmic strings in an expanding universe, provided by Paul Shellard.*[29]

energy can act as seeds around which fluctuations in density can start to develop and from which ultimately galaxies might form. However, it is very difficult to calculate what would happen in detail. A host of complicated processes come into play and the fastest computers in the world are still unable to follow all these processes quickly and accurately enough to determine whether strings can produce real galaxies clustered in the patterns that we see. The acid test of such a theory is again provided by the pattern of fluctuations in the microwave radiation left over from the Big Bang. The gravitational field created by the evolving network of strings will leave its characteristic imprint in this radiation. It appears to have a signature that is quite different from that left by the inflated zero-point fluctuations which provide the rival theory. But not everyone agrees. So far, if the string predictions have been correctly calculated, the evidence of the ground-based detectors is beginning to turn against them, but it is early days. The predictions need to be more fully worked out by bigger computer simulations and elaborated, and only the satellite observations will be fully convincing checks.

The cosmic string scenario for the origin of galaxies is a natural rival to the inflationary theory. In the string theory, the initial non-uniformities in the density of the Universe from place to place are created by the appearance of string loops in different places, from the definite continuous vacuum structure of particular energy fields, whereas in the inflationary theory they arise from the zero-point fluctuations. The two ideas are not natural bed-fellows. Just as inflation sweeps away walls and monopoles that have formed in the Universe so it will sweep away the distribution of cosmic strings if they form before inflation happens. In that case they will play no further role in the formation of galaxies. Thus, if galaxies owe their existence to a population of cosmic strings forming in the very early Universe, either inflation did not happen, and we cannot appeal to any of its other benefits to explain mysterious properties of the Universe, like its proximity to the critical rate of expansion today, or the vacuum structure of the ultimate unified theory has a very peculiar double structure. That structure must undergo a slow change that first enables inflation to occur and then be followed by a further particular type of change which permits cosmic strings to appear without any walls or monopoles appearing along with them. Most cosmologists think that this is a tall order and rather unlikely. However, there is no proof of its impossibility.

Individual cosmic vacuum strings are strange beasts. They could reveal their presence by bending rays of light that move close to them. The defining characteristic of a string is its mass per unit length. The larger this is, the greater will be the mass and gravitational effect of any piece of string on other masses. If a straight line of cosmic string were to pass through this page then its effect on neighbouring masses would be to make them move together. It is as if a wedge is cut out of space around the string and the remaining space pulled together to fill the gap (see Figure 8.19).

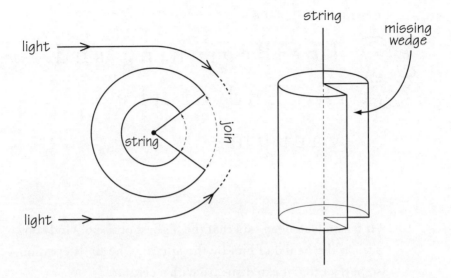

Figure 8.19 *A long cosmic string passing into the page has the same effect as removing a wedge of space around the string. This creates a focusing of light rays as they pass by the string as if they are passing through a lens.*

Strings would be like nothing else we have ever encountered. If a piece of vacuum string extended across a part of space that astronomers were observing then the effect of its gravity would be to behave like a lens. A star lying behind the string would have its image duplicated.[30] A curving piece of string would create a tell-tale line of double images. Astronomers have looked for these tell-tale images but have yet to find them. Plenty of multiple images have been seen by the Hubble Space Telescope and they are clearly due to the lensing action of gravity fields. However, they seem to be caused by very large intervening objects like galaxies not cosmic strings.

These speculative possibilities show some of the unending richness of the physicists' conception of the vacuum. It is the basis of our most successful theory of the Universe and why it has the properties that it does. Vacuums can change; vacuums can fluctuate; vacuums can have strange symmetries, strange geographies, strange histories. More and more of the remarkable features of the Universe we observe around us seem to be reflections of these properties of the vacuum. All that remains for us to ask about it is whether it had a beginning and whether it will ever have an end.

chapter nine

The Beginning and the End of the Vacuum

"It has indeed been said that the highest praise of God consists in the denial of him by the atheist, who finds creation so perfect that it can dispense with a creator."

Marcel Proust[1]

BEING OUT OF NOTHINGNESS

"The human race, to which so many of my readers belong, has been playing at child's games from the beginning . . . The players listen very carefully and respectfully to all that the clever people have to say about what is to happen in the next generation. The players then wait until all the clever men are dead and bury them nicely. They then go and do something else. That is all. For a race of simple tastes, however, it is great fun."

G.K. Chesterton[2]

Why is there something rather than nothing? Some regard such questions as unanswerable, some go further to claim that they are meaningless, whilst others claim to provide the answers. Science has proved a reasonably effective way of finding out about the world because it confines itself, in the main, to questions about 'how' things happen. If it does ask the question 'why' it is generally about an aspect of things that can be answered if one is in possession of a full description of how a certain sequence of events occurs, what causes what, and so on. As one digs deeper to the roots of scientific theories one finds that there is a foundation of a sort that we call laws of Nature, which govern the behaviour of the most elementary particles of Nature. The identities of these particles, the things they are able to do, and the ways in which they can combine are like axioms whose consequences we can test against the facts of experience. To some extent we may find that it is very difficult to imagine how things could be otherwise because the properties of the laws become closely bound up with the nature of the populations of identical elementary particles that they govern. Some laws only act upon particles with particular attributes. But in other respects it is possible for us to envisage a universe that was slightly different from our own. So far, we have not found a theory that requires there to be only one possible universe. This question boils down to one about the nature of the vacuum landscape for the ultimate theory of the Universe. If there is a single valley in this landscape, then there is a single possible vacuum state and one possible set of values for the constants of Nature that define it. If there are many valleys, and so many vacua, then the constants of Nature are not uniquely specified by one possibility. They can exist with different values and, as we have seen in the last chapter, they may even do so elsewhere in our Universe now. Hence there has emerged a more modest version of the great ontological question, 'Why is there something rather than nothing?' which physicists are able to comment on in a meaningful way. From their perspective, certain aspects of the world may be inevitable or be necessary features of any universe that is going to contain living observers.

The matters of science which are relevant to our great question are those studied by cosmologists and physicists. The study of the Universe has revealed it to be expanding. Tracing its history backwards for billions of years leads us to a moment when densities and temperatures would become infinite, and further backtracking using this description is impossible. This leads us to consider the serious possibility that it may have had a beginning at a finite time in the past. This is only an extrapolation and needs to be examined far more closely if it is to be taken seriously, but let us for the moment take it seriously enough to follow the argument a little further. If the expansion did have a beginning then we are faced with further questions: is this 'beginning' merely the start of the expansion of the Universe that we see today or is it the Beginning, in every sense, of the entire physical Universe? And, if it is the latter, does it include just the matter and energy in the Universe, or the entire fabric of space and time as well? And, if space and time come into being, what of the laws and symmetries and constants of Nature, as we like to call them; do they appear as well? Lastly, if some or all of these things must come into being at some identifiable moment of history, what do they come into being from, why, and how?

There are ancient traditions of humans asking great 'why' questions about the nature and end of existence. A large fraction of the readers of these pages will live in societies that have been strongly influenced by the Judaeo-Christian tradition and the ideas that it generated in order to harmonise its writings and doctrines with our early knowledge of the physical world. The doctrine of the Universe having been created out of nothing (*creatio ex nihilo*) is almost unique to the Christian traditions. A survey of the mythological beliefs of the world reveals surprisingly few basic cosmological scenarios, despite a veneer of exotic actors and fantastic memorable mechanisms.[3]

The idea of a 'created' universe is most commonly found as a re-shaped or reconstructed universe, usually being fashioned out of a state of chaos or a structureless void. Alternatively, the world may 'emerge' from some other state, for which there are a multitude of candidates. It can spring new-born from some primeval womb, or be fished out of the dark waters of chaos by a heroic diver. It may hatch from a pre-existent egg or emerge from the union of two world parents. Elsewhere, we find versions of the story of a clash between some superhero and the forces of darkness and evil, out of which the world is born. All these pictures have close links to human experiences of human childbirth, battles with rivals, animal reproduction and fishing for food. The emergence of something from nothing, like the birth of a child, is accompanied by pain and effort. It is often opposed, but ultimately it succeeds. Not all these examples are straightforward. As time passes mythological accounts tend to become increasingly complex. More and more facts come to light about the world and new questions are asked. Answers can usually only be provided by embroidering the story further. Explanations grow more elaborate.

Other traditions can be found in which the Universe does not begin at all: it always was. These traditions often have a cyclic picture of time and history that owes much to the seasonal cycles exploited by agricultural societies and the cycles of human life and death.[4] Thus, while the ultimate reality continues from a past eternity to a future eternity, the Earthly world will die only to be reborn, rising like a phoenix from the ashes of its predecessor. The pattern of human cosmological stories is summarised in Figure 9.1.

In these accounts the language of creation, in the artistic or practical sense, is often used to describe the bringing into being of the observed state of the world. In most cases the raw materials were given and the creative process fashioned order from chaos. No inquiry was made into the origin of the materials themselves. The origin of the world out of the union of two gods offers scope for contemplating the appearance of something where once there was nothing, in the same way that a new child is not just a rearrangement of pre-existing things. However, the idea of making something from nothing was compromised by the pre-existence of the gods. The offspring owed something to them just as the offspring of a human union displays characteristics of its parents. These stories always drew a veil over the question of how there could be a transition from absolutely nothing to something. No tradition addresses this question. All have something emerging from something else, usually aided by an act of will by a superhuman intermediary. The impression that one gets from these stories is that the idea that the world began was not too difficult to accept, but it was impossible to comprehend the idea that it could have 'begun' in any sense other than having changed from something else into what it now appears to be. Nothing, as we have seen, is a very difficult concept to grasp. Here, it was easy to sidestep it.

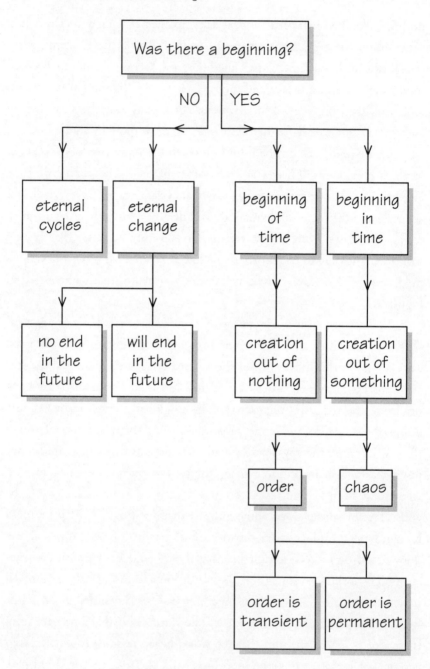

Figure 9.1 *The common patterns of cosmological traditions.*

CREATION OUT OF NOTHING

"In the beginning there was Aristotle,

And objects at rest tended to remain at rest,

And objects in motion tended to come to rest,

And soon everything was at rest,

And God saw that it was boring."

Tim Joseph[5]

There is a popular notion that the Christian tradition of the creation of the Universe out of nothing is simply that God made the Universe appear out of nothing at a moment in the finite past. Everything that constitutes the world – space, time, matter, laws of Nature – sprung into being at once out of nothing at all. These things were not fashioned out of some simpler, less ordered or chaotic mess. They were *created*, not *formed* out of something else.

Almost all of the statements in the last paragraph have a number of different variants and interpretations. Yet the detailed nature of the traditional doctrine of creation out of nothing is far less specific and cosmological. One suspects that the religious ideas have gradually become far more specific and well defined with the advent of twentieth-century cosmology and the fairly precise picture that it gives of the expanding Universe and its apparent beginning. Although some modern theologians seek to reconcile the ancient tradition of creation out of nothing with contemporary cosmological ideas,[6] it is good to recall that the doctrine of creation out of nothing did not arise in Christian tradition in order to make assertions about astronomy and cosmology as we now understand them. Its primary objectives were to make a statement about the relationship between God and the Universe; to assert that there was meaning and purpose to the world, that it was dependent upon God, and to distinguish clearly between Christian beliefs and those of other belief systems that were current at the time when the early Church was developing its theology.[7] In particular, it proclaimed that Nature was not the same thing as God; this was an important distinction that served to make the worship of idols or nature-gods appear futile. It also sought to make a theological point about the power of God. Creation rather than formation out of pre-existing stuff asserted that no help was needed from other sources; God controlled matter's existence as well as its orderly patterns.

These aims were of far greater significance than any desire to underpin the idea that the world came into being at a definite moment in time.[8] Yet it is this latter idea that often seems to be of primary interest to many modern apologists trying to reconcile science and Christianity.[9] This polemical use of a doctrine of creation out of nothing is certainly not new. It was first developed in order to distinguish its adherents from other philosophies and beliefs, both current and inherited, in the Graeco-Roman worlds. Thus, whereas the followers of Plato subscribed to the shaping of some pre-existent world of matter into its present form by the action of a demiurge, Christianity affirmed a creation out of no pre-existent material at all. Aristotle, by contrast, argued for the past eternity of the world rather than its sudden appearance.

One might imagine that early Christianity inherited the idea of creation out of nothing from Judaism, but the evidence for its presence in Jewish texts is by no means incontrovertible or unambiguous. There is no explicit statement in early Jewish writings about the creation of the Universe. It was not a doctrinal ingredient of its theology. There appears to have been little interest in the question. There was no doubt in anyone's mind about the existence and omnipotence of Yahweh and so no motivation for proofs of His existence that appealed to the need for a Creator. Although no systematic position is explicitly worked out there is evidence that, if pressed on the issue, there was a clear defence of the concept of creation out of nothing. At the end of the first century, the influential Rabbi Gamaliel engaged in a debate[10] with a philosopher on the question of whether the world was formed from pre-existent materials. The philosopher describes the work of God as that of a great artist using the materials ('colours') that are made available in the opening verses of Genesis. But Gamaliel counters this theory by arguing that all these 'colours' are explicitly described in the Bible as having been created by God. Thus he rejects an interpretation of the first two verses of Genesis as supporting the idea that the world was formed out of pre-existing material. Implicitly, he asserts creation out of nothing by declaring that anything you nominate as the raw material for creation was created by God. Here, as in other places, we seem to be seeing a theological affirmation that happens to use cosmological categories rather than the development of any explicit cosmological theory that can be used to deduce other properties of the Universe. There would have been religious views about the end of the world but they would not in any sense have been consequences of the teachings about its beginning.

The Wisdom of Solomon[11] speaks of the formation of the world out of formless matter that the Almighty 'made'. The text most frequently cited as the earliest explicit statement of the idea of creation from nothing is in the second book of Maccabees,[12] which speaks of the 'Creator of the world' bringing about 'the beginning of all things'. The context is a story in which a mother of seven martyrs encourages her youngest son to remain faithful by calling upon him to 'look upon the heaven and the earth, and all that is therein, and consider that God made them of things that were not' and be assured that ultimately he will awaken the righteous from death. But there is no philosophical purpose in the mother's mind. She is just basing hope for resurrection upon her faith in the power of God. Other examples can be found of similar phraseology being used to express the coming of children into the world 'out of non-being'.[13] Again, there is no engagement with the sort of tricky philosophical problems for which the possibility of creation out of nothing would be a possible remedy or counter-example.

In the earliest Christian traditions there is, accordingly, no ready-made inherited position about the creation of the world out of nothing. There is considerable freedom to develop this idea gradually during the first and second centuries, for nowhere in the New Testament writings is the doctrine of creation out of nothing explicitly taught. It began to be discussed seriously by theologians in about AD 160 as a result of the challenging questions raised by Gnostic philosophies.

In Gnosticism the questions of 'why' and 'how' the world was created were of great significance, not because the Gnostics were especially interested in cosmology but because of their negative view of the world. They needed to have some explanation as to why this defective, immoral world came into being, and how it could result from the actions of the one true and perfect God. Gnostics maintained that the world was the creation of a group of more limited lesser beings ('angels') who either did not know the true God or were in rebellion against Him. They viewed matter and the physical Universe as something possessing only a partial reality which disturbed the true plan for the Universe. The ensuing process of salvation had as its primary goal the destruction of the defective material world. It was the complex evolution of the debate between the Gnostics and their opponents (and a whole spectrum of intermediate positions) in the early Church that led to the emergence in the early Christian Church of a clear doctrine of the creation of the Universe out of nothing in the writings of Basilides, Valentius and Irenaeus.

Basilides and his school in Antioch developed a Gnostic system unlike all others. It focused on the need to determine the nature of creation itself. Basilides proposed that in the beginning there was just pure ineffable Nothing.[14] It may be that he equated Nothing with God, and on one occasion he describes God as 'non-being'. This is probably just a rather extreme use of a form of negative theology in which one defines God in terms of the things that He is not.[15] Unlike other Gnostics, Basilides rejects the idea that there is some germinating world-seed or pre-existent formless matter from which the world emerges. He regarded such devices as limitations on the power and superhuman nature of God. He rejects totally the idea that God works like a human craftsman or an artist using the materials that are at hand to fashion the Universe.

This is the earliest explicit rejection of the general idea of the formation of the world out of formless materials. From now on it became clear that divine creation must be placed on a higher plane than artistic creation.[16] Basilides' views became widely accepted and the rejection of the formation model for the origin of the world allowed the idea of creation out of nothing to become established during the second half of the second century. Quite quickly, the world-formation model came to be regarded as impossible to reconcile with the biblical concept of creation. Previously, the concept of 'nothing' was often defined in such a way that a formation out of pre-existent material could be accommodated within a statement of creation out of nothing, but Basilides, a non-standard Gnostic, was the first Christian theologian[17] to speak unambiguously about creation out of nothing in a very inflexible sense that was designed to be exclusive.

In less than a generation, a surprising change of attitude had occurred. In the middle of the second century, the early Christian Church had no interest in any specific doctrine of the creation of the world and would have been happy to accommodate a picture of the world forming out of pre-existent material with the Genesis account. Basilides' careful argument turned things around. Creation *ex nihilo* was adopted as a central doctrine and the theories of world formation out of anything other than nothing were rejected as heretical challenges to the omnipotence of God and an adherence to the heretical theories of the godless philosophers. The resulting doctrine emerges from a synthesis of three convictions: that creation occurs 'out of nothing', that God is the supreme Creator, and the rejection of the tempting old idea that God acts in a way that is analogous to human creative action.

It is curious that the Christian doctrine of creation out of nothing was introduced by a Gnostic, since the doctrine is by no means a Gnostic idea. Its Gnostic legacy is a reflection of the more sophisticated cosmological thinking that the Gnostics developed in order to deal with their own complicated doctrinal problems. They thought that the version of Christian truth that they taught was plainly superior to the insights coming from existing philosophy and science.

The rejection of the world-formation cosmology was first made explicit in the works of Tatian and Theophilus of Antioch (Basilides' home town also), but there would later emerge a view that formless matter was created out of nothing and then shaped into an orderly universe. Tatian claimed that matter is made out of nothing by God and Theophilus developed a solid biblical basis for the doctrine of creation out of nothing.

From the modern perspective it is easy to wonder why early theologians seem to make such heavy weather of all this. There seem to be so few alternatives to the creation-out-of-nothing idea and it seems strange that such a complicated sequence of events was needed for the alternatives to be mapped out clearly. It is important to remember that one reason for their slowness is simply that they were not looking for such a doctrine. They were not motivated by a special interest in astronomy or natural philosophy. Parts of their doctrine were constructed occasionally when needed to defend specific theological points. It was synthesised into a fully worked-out form only when it was needed to counter the theological consequences of rival Greek views about the world being fashioned from pre-existent matter. Creation out of Nothing is one of the by-products of the early Christian Church's disputes with the ideas of Greek philosophy.

One must also remember the confusing background of Platonic philosophical ideas which were still very influential. The Platonic view of the world was that there exists an unseen eternal realm of ideal 'forms' which are the perfect blueprints of the things that we see in the material world. Thus each triangular shape that we see drawn on a piece of paper is an imperfect representation of the ideal triangular form. This makes the idea of nothing a very difficult one to entertain. Even if you wish to conceive of a moment before which the material world did not exist, the eternal forms still exist. Complete Nothingness is inconceivable. Thus the world-formation cosmologies which produce order in chaotic or unformed material can be seen as *in-forming* the raw material with the patterns from the eternal forms — transferring 'information' content as we might say today. In modern approaches to these problems the Platonic worry still exists in a slightly different form. We can perhaps imagine that no material universe exists, maybe even that no laws of Nature exist, but nothing *at all* is unimaginable for us because it would mean no facts could exist — not even a fact like the statement that nothing exists, in fact.

PHILOSOPHICAL PROBLEMS ABOUT NOTHING AND HOW WE ESCAPED FROM IT

"Every public action, which is not customary, either is wrong, or, if it is right, is a dangerous precedent. It follows that nothing should ever be done for the first time."

Francis Cornford[18]

The question of why there is a world at all was raised in a short pamphlet by the philosopher Leibniz in 1697 entitled 'On the Ultimate Origination of Things'.[19] Leibniz realised that it did not matter whether you thought the world was eternal or appeared out of nothing as maintained by orthodox Christian doctrine. All theories and beliefs still faced the problem of why there was something rather than nothing. Philosophers took little interest in this question for a long time after Leibniz. Problems like this were not part of an analytical philosophy that built up understanding of things step by step. Leibniz's problem needed an understanding of everything all at once. It was too ambitious. In fact, it was as good a candidate as any for an intrinsically insoluble problem.[20] Philosophers who considered the question, like Wittgenstein ('Not *how* the world is, is the mystical, but *that* it is')[21] and Heidegger, had little to say in answer to it and appear more interested in wondering about why the question is one that we find so compelling.[22]

The only novel contribution to this problem before the twentieth century was the consideration of whether the well-defined concept of mathematical existence had any cosmological implications. The development of axiomatic mathematical systems, in which a system of self-consistent rules ('axioms') were laid down and consequences deduced or constructed from them, led to a 'creation' of mathematical truths that 'existed' in a rather particular sense. Any mathematical statement that was logically consistent was said to 'exist'. Mathematicians would produce what became known as 'existence proofs'. This is clearly a far broader concept of existence than physical existence. Not all the things that are logically possible seem to be physically possible and not all of those now seem physically to exist. However, a philosopher like Henri Bergson clearly thought that this type of weak mathematical existence was a possible avenue along which to search for a satisfying solution to Leibniz's problem:[23]

> "I want to know why the universe exists . . . Whence comes it, and how can it be understood, that anything exists? . . . Now, if I push these questions aside and go straight to what hides behind them, this is what I find: – Existence appears to me like a conquest over nought . . . If I ask myself why bodies or minds exist rather than nothing, I find no answer; but that a logical principle, such as A = A, should have the power of creating itself, triumphing over the nought throughout eternity, seems to be natural . . . Suppose, then, that the principle on which all things rest, and which all things manifest, possesses an existence of the same nature as that of the definition of the circle, or as that of the axiom A = A: the mystery of existence vanishes . . ."

Unfortunately, this approach to why we see what we see is doomed to failure. As the nature of axiomatic systems has become more fully appreciated it is clear that *any* statement can be 'true' in some mathematical system. Indeed, a statement which is true in one system might be false in another.[24]

As an interesting sidelight, there is an amusing dialogue reproduced in Andrew Hodges' biography[25] of Alan Turing. Turing attended Wittgenstein's lectures on the philosophy of mathematics in Cambridge in 1939 and disagreed strongly with a line of argument that Wittgenstein was pursuing which wanted to allow contradictions to exist in mathematical systems. Wittgenstein argues that he can see why people don't like contradictions outside of mathematics but cannot see what harm they do inside mathematics. Turing is exasperated and points out that such contradictions inside mathematics will lead to disasters outside mathematics: bridges will fall down. Only if there are no applications will the consequences of contradictions be innocuous. Turing eventually gave up attending these lectures. His despair is understandable. The inclusion of just one contradiction (like $0 = 1$) in an axiomatic system allows *any* statement about the objects in the system to be proved true (and also proved false). When Bertrand Russell[26] pointed this out in a lecture he was once challenged by a heckler demanding that he show how the questioner could be proved to be the Pope if $2 + 2 = 5$. Russell replied immediately that 'if twice 2 is 5, then 4 is 5, subtract 3; then $1 = 2$. But you and the Pope are 2; therefore you and the Pope are 1'! A contradictory statement is the ultimate Trojan horse.

This temptation to replace physical existence by mathematical existence can be taken to extremes. Suppose that we imagine that all possible mathematical formalisms are laid out in front of us. They each appear like a great network of all possible deductions that follow from their axioms. If the mathematical system is very simple then the deductions will also be very limited in their complexity. But if the axioms are rich enough then the sea of deductions will include extremely complex structures which possess the capability of self-awareness. It is as if we are building a computer simulation of how a system of planets might form around a star. We tell the computer all the laws of motion and gravity, and whatever other physics and chemistry that we want included in the story. The computer will produce a simulation, or artificial sequence of events, culminating in the formation, say, of a planet like the Earth. We could imagine a future in which the computational capability was such that the simulation could be continued in great detail. Biochemical replication could be followed and early life forms simulated. Eventually, the complexity of the replicators being modelled in the computer could reach a level that displayed self-awareness and an ability to communicate with other self-aware sub-processors in the simulation. They might even engage in philosophical debates about the nature of the simulation, whether it was designed for them, and whether there exists a Great Programmer behind the scenes. At root these 'conscious' sub-programmes would exist only in the logical structure of the computer.

They would be part of the mathematical formalism being explored and elaborated by the machine.

We can ask whether the possibility of containing structures able to be self-aware is a general or a rather special property of mathematical formalisms. One day it may be possible to answer this but at present we can only make rather weak statements. There have been controversial proposals[27] that the Gödel incompleteness[28] properties of arithmetic may be necessary for consciousness to operate as it does in humans. If true, this would be equivalent to saying that mathematical systems need to be rich enough to contain arithmetic in order to contain structures with the complexity of human consciousness. Thus, Euclidean geometry, which is smaller than arithmetic and does not possess incompleteness, would be too simple a logical system to become self-aware. If this approach could be developed further then we might be able to isolate a collection of mathematical structures which allow the possibility of encoding conscious sub-programmes. Conscious life would 'exist' in the mathematical sense only in these mathematical formalisms.

Most philosophers treat such recipes with distaste. They regard real physical existence as distinct from mathematical existence. In the words of Nicholas Rescher,[29]

> ". . . getting real existence from pure logic is just too much of a conjuring trick. That sort of hat cannot contain rabbits."

Mathematical existence allows anything to 'exist'. Some axiomatic system can always be framed in which any statement is true (and others found in which it is false). This type of existence does not, therefore, really explain anything. We want to know why so much of what we see around us can be explained as a truth of a particular system of logical rules with a single set of axioms. The fact that those axioms are not too exotic shows that the world can be described by quite simple ideas (that is, ones that are intelligible to human beings) to a very surprising degree.

CREATION OUT OF NOTHING IN MODERN COSMOLOGY

"Then God created Bohr,
And there was the principle
And the principle was quantum,
And all things were quantified,
But some things were still relative
And God saw that it was confusing."

Tim Joseph[30]

The discovery of the general theory of relativity by Einstein enabled the first mathematical descriptions of entire universes to be made. Only very simple solutions of Einstein's equations have been found completely by direct calculation, but fortunately these simple solutions are extremely good descriptions of the visible part of the Universe for a considerable time in the past. They describe expanding universes in which the distant clusters of galaxies are moving away from each other at ever-increasing speeds. Deviations from the exact symmetry of the special solutions can be introduced quite easily, so long as they are small, and this results in a good description of the real non-uniformities in the Universe.

As we try to reconstruct the past history of these cosmologies, we encounter a striking feature. If matter and radiation continue to behave as they do today, and Einstein's theory continues to hold, then there will be a past time when the expansion must have encountered a state of infinite density and temperature. When this property was first appreciated, it sparked a number of very different reactions. Einstein[31] thought that it was merely a consequence of considering expanding universes that contained matter without significant pressure. If pressure was included then he thought that it would resist the contraction of a universe down to infinite density, just as air pressure resists our attempts to squeeze an inflated balloon into a very small volume. It would 'bounce' back. But this intuition was completely wrong. When normal pressures were included in the universe models it made the singularity *worse* because in Einstein's theory all forms of energy, including those associated with pressures, have mass and gravitate by curving space. The singular state of infinite density remained. Others objected that the singular 'beginning' only appeared because we were looking at descriptions of expanding universes which were spherical, with expansion at exactly the same rate in every direction. If the rate was made slightly different in different directions then, when the expansion was retraced backwards in time, the material would not all end up in the same place at the same time and the singularity would be avoided. Unfortunately, this also proved to be no defence against the singular beginning. Rotating, asymmetrical, non-uniform universes all had the same feature: an apparent beginning. If matter was present in the universe, its density was infinite there.

The next attempts to evade this conclusion looked to a more subtle possibility. Perhaps it was just the way of measuring time and mapping space in the model universe that degenerated into a singularity, just as with the coordinates on a globe of the Earth's surface. At the Poles the meridians intersect and create a singularity in the mapping coordinates; yet nothing odd happens on the Earth's surface. Likewise, perhaps nothing dramatic happens at the Universe's apparent beginning; you merely change to measuring time and space in a new way and repeat this process, as required, indefinitely into the past.

These possibilities created considerable uncertainty for cosmologists until the mid-1960s. They were removed by an approach pioneered by Roger Penrose.[32] He looked at the problem in a new way and considered the collection of all possible histories that were possible for all particles of matter and light rays. Bypassing all the problems of the shape and uniformity of the Universe and the ways of measuring time, Penrose showed that if Einstein's theory is true, if time-travel is impossible, and gravity is always attractive, then so long as there is enough gravitating matter and radiation in the Universe, at least one of that collection of histories must have had a beginning – it cannot be continued indefinitely into the past. Observations showed that there was easily enough matter to meet the last condition[33] and all forms of matter then known or hypothesised exhibited gravitational attraction.

This deduction was remarkable in many respects. It managed to come up with such a strong and general conclusion because it gave up the idea that it was the infinite density – the 'Big Bang' itself – that characterised the beginning of a universe. Instead, it employed the simpler and more relevant idea of a history with a beginning – that the universe of space and time had an edge. It might be that the histories with a beginning are accompanied by infinite densities but that is a quite separate, and much more difficult, question which is still not fully answered.[34] Also, it is only demanded that *one* past history have a beginning, not all of them. The simple expanding universes which describe our Universe so well today have the property that all the histories come to an end simultaneously at a finite time in the past when the density becomes infinite. Penrose's approach tells us nothing about the nature of the beginning of the histories, only that they must occur when the assumptions he makes hold good.

The interesting thing about the singularity that is predicted by these theorems is that there is no explanation as to *why* it occurs. It marks the edge of the Universe in time (see Figure 9.2). There is no before; no reason why the histories begin; no cause of the Universe. It is a description of a true creation out of nothing.

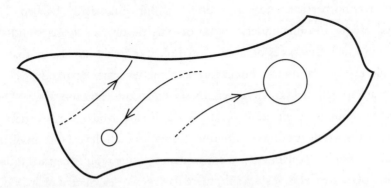

Figure 9.2 *Singularities are part of the edge of space and time. If we represent space-time as a sheet then this edge can be at places where the density of matter becomes infinite or even places where it remains finite because there are 'holes' in the sheet.*

These developments led to considerable interest amongst theologians and philosophers of science,[35] who saw it as a demonstration that the Universe did have a beginning in time. From the mid-1960s until about 1978 these mathematical theorems were widely cited as evidence that the Universe had a beginning. However, it is important to realise that they are mathematical *theorems* not cosmological *theories*. The conclusions follow by logical deduction from the assumptions. What are those assumptions and should we believe them? Unfortunately, the two central assumptions are now not regarded as likely to hold good. We expect Einstein's equations of general relativity to be superseded by an improved theory that successfully includes the quantum effects of gravitation. This new theory will have the property of becoming just like Einstein's theory when densities are low, as they are now in the Universe. Indeed, recent superstring theories of elementary particles and gravity, which are the favourite candidates for an ultimate theory of all the forces of Nature, have the nice property of reducing to Einstein's equations in a low-energy environment. It is widely expected that this new improved theory will not contain the singular histories that characterised Einstein's theory, but until we have the new theory we cannot be sure.

There is a more straightforward objection to the deduction of a beginning using the theorems of Penrose and Hawking. The central assumption is that gravity is always an attractive force. When the theorems were first proved this was regarded as an extremely sound assumption and there was no particular reason to doubt it. But things have changed. The rapid progress in our understanding of particle physics theories and the ways in which the forces of Nature are linked together has shown that we should expect Nature to contain forms of matter which respond *repulsively* to gravitational fields. Moreover, these fields are very appealing. They include amongst their number the scalar fields which drive inflation. Indeed, the whole process of inflation, through which the expansion of the Universe can be accelerated, is a consequence of the repulsive gravitational action of these fields. As a result, there has been a sea change in attitudes. Whereas up until the late 1970s it was widely accepted that all matter in the Universe should exhibit gravitational attraction and the assumptions of the singularity theorems hold good, since 1981 exactly the opposite has been believed: that it is unlikely and undesirable that all matter displays gravitational attraction. Indeed, the recent observations of the acceleration in the expansion of the Universe today, if correct, demonstrate that there exists matter which displays gravitational repulsion. It is the cosmological vacuum energy that contributes a repulsive lambda force to the gravitational force of Newton.[36]

The logic of the singularity theorems is that if their assumptions hold then there must be a singularity in the past. If the assumptions do not hold, as we now believe is most likely, then we cannot conclude that there is no beginning – only that there is no theorem. Some universes with gravitationally repulsive matter still have beginnings where the density is infinite, but they don't need to. We have already seen one spectacular example that appears to evade the need for a beginning. The self-reproducing eternal inflationary universe almost certainly has no beginning. It can be continued indefinitely into the past.

Thus the old conclusions of the singularity theorems are no longer regarded by cosmologists as likely to be of relevance to our Universe. Crucial assumptions in those theorems – the attractive nature of gravitation, and the truth of Einstein's general theory of relativity all the way back to the earliest times when energies are so high that quantum gravitational effects must intervene – are no longer likely to be true. What are the alternatives?

NO CREATION OUT OF ANYTHING?

"We are the music-makers
And we are the dreamers of dreams,
Wandering by lone sea-breakers,
And sitting by desolate streams;
World-losers and world-forsakers,
On whom the pale moon gleams:
Yet we are the movers and shakers
Of the world forever, it seems."

Arthur O'Shaughnessy, 'Ode'

If the whole expanding Universe of stars and galaxies did not appear spontaneously out of nothing at all, then from what might it have arisen? One option that has an ancient pedigree is that it had no beginning. It has always existed. A persistently compelling picture of this sort is one in which the Universe undergoes a cyclic history, periodically disappearing in a great conflagration before reappearing phoenix-like from the ashes.[37] This scenario has a counterpart in modern cosmological models of the expanding universe. If we consider closed universes which have an expansion history that expands to a maximum and then contracts back to zero (see Figure 9.3), then there is a tantalising possibility. Here, we see a one-cycle universe that begins at a singularity and ends at one.[38] But suppose the Universe re-expands and repeats this behaviour over and over again. If this can happen then there is no reason why we should be in the first cycle. We could imagine an infinite number of past oscillations and a similar number to come in the future. We are ignoring the fact that a singularity arises at the start and the end of each cycle. It could be that repulsive gravity stops the Universe just short of the point of infinite density or some more exotic passage occurs 'through' the singularity, but this is pure speculation.

This speculation is not entirely unrestrained, though. Let us assume that one of the central principles governing Nature, the second law of

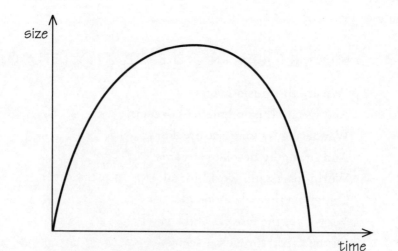

Figure 9.3 *A one-cycle closed universe.*

thermodynamics, which tells us that the total entropy (or disorder) of a closed system can never decrease, governs the evolution from cycle to cycle.[39] Gradually, ordered forms of matter will be transformed into disordered radiation and the entropy of the radiation will steadily increase. The result is to increase the total pressure exerted by the matter and radiation in the Universe and so increase the size of the Universe at each successive maximum point of expansion,[40] as shown in Figure 9.4. As the cycles unfold they get bigger and bigger! Intriguingly, the Universe expands closer and closer to the critical state of flatness that we saw as a consequence of inflation. If we follow it backwards in time through smaller and smaller cycles it need never have had a beginning at any finite past time although life can only exist after the cycles get big enough and old enough for atoms and biological elements to form.

For a long time this sequence of events used to be taken as evidence that the Universe had not undergone an infinite sequence of past oscillations because the build-up of entropy would eventually make the existence of stars and life impossible[41] and the number of photons that we measure on average in the Universe for every proton (about one billion) gives a measure of how much entropy production there could have been. However, we now know that this measure does not need to keep on increasing from

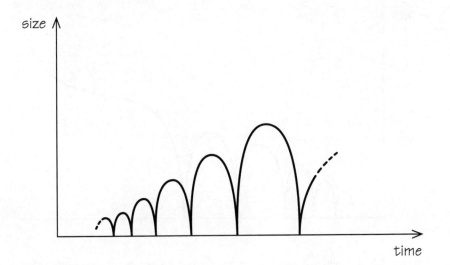

Figure 9.4 *A many-cycle closed universe in which the cycles increase in size.*

cycle to cycle. It is not a gauge of the increasing entropy. Everything goes
into the mixer when the Universe bounces and then the relative number of
protons and photons gets set by physical processes that occur early on. One
problem of this sort might be that of black holes. Once large black holes
form, like those observed at the centres of many galaxies, including the
Milky Way, they will tend to accumulate in the Universe from cycle to
cycle, getting ever more massive until they engulf the Universe, unless they
can be destroyed at each bounce or become separate 'universes' which we
can neither see nor feel gravitationally.

A curious postscript to the story of cyclic universes was recently discovered by Mariusz Dąbrowski and myself. We showed that if Einstein's
lambda force does exist then, no matter how small a positive value it takes,
its repulsive gravitational effect will eventually cause the oscillations of a
cyclic universe to cease. The oscillations get bigger and bigger until eventually the Universe becomes large enough for the lambda force to dominate
over the gravity of matter. When it does so, it launches the Universe off
into a phase of accelerating expansion from which it can never escape unless
the vacuum energy creating the lambda stress were to decay away mysteri-
ously in the far future (see Figure 9.5). Thus the bouncing Universe can

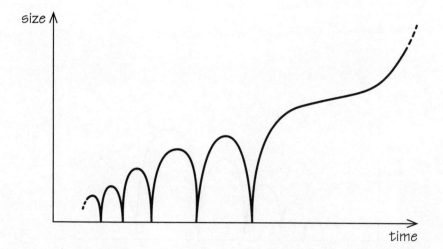

Figure 9.5 *A many-cycle universe is eventually transformed into an expanding universe
by the presence of a lambda force.*

eventually escape from its infinite oscillatory future. If there has been a past eternity of oscillations we might expect to find ourselves in the last ever-expanding cycle so long as it is one that permits life to evolve and persist.

Another means by which the Universe can avoid having a beginning is to undergo the exotic sequence of evolutionary steps created by the eternal inflationary history that we explored in the last chapter. There seems to be no reason why the sequence of inflations that arise from within already inflating domains should ever have had an overall beginning. It is possible for any particular domain to have a history that has a definite beginning in an inflationary quantum event, but the process as a whole could just go on in a steady fashion for all eternity, past and present.

One of the most interesting features of research efforts in modern cosmology is the way in which the creation-out-of-nothing tradition influences the direction in which cosmologists look for mathematical models of the early Universe. The singularities predicted by the theorems of Hawking and Penrose in the 1970s were happily accepted by many cosmologists as a real prediction of Einstein's theory of gravitation, even though they were really just predicting that the theory had to cease being a good descriptor of the Universe at some finite time in the past, when densities became too high for the quantum effects of gravity to be ignored any longer. In other areas of physics, the appearance of predictions that physically measurable quantities become infinite is always a signal that the theory has ceased to be applicable to the circumstance to which it is being applied. A refinement is necessary to make the equations applicable to a wider range of physical phenomena. Yet the appearance of an infinity in the density of matter and a beginning to space and time was regarded as acceptable to many scientists. The breakdown of prediction was often interpreted as a consequence of the Universe having a beginning rather than as an incompleteness of the theory. This is perhaps because the picture created by having a 'beginning' for the Universe is one with which most Westerners feel comfortable because of the religious traditions in which they have been raised.

For similar reasons, there often seems to be more opposition to the idea of a universe that has always existed. The steady-state cosmology of

Herman Bondi, Fred Hoyle and Thomas Gold attracted much opposition from scientists and non-scientists alike. That opposition came from opposite ends of the religious spectrum. Some Christians opposed it because it denied the reality of sudden creation out of nothing whilst the Stalinist regime in the Soviet Union disliked it because it denied the possibility of progress and evolution towards a better world.

At first, the absence of a beginning appears to be an advantage to the scientific approach. There are no awkward starting conditions to deduce or explain. But this is an illusion. We still have to explain why the Universe took on particular properties – its rate of expansion, density, and so forth – at an *infinite* time in the past.

There are several specific candidates for the something out of which the present expansion of the Universe might have emerged. Figure 9.6

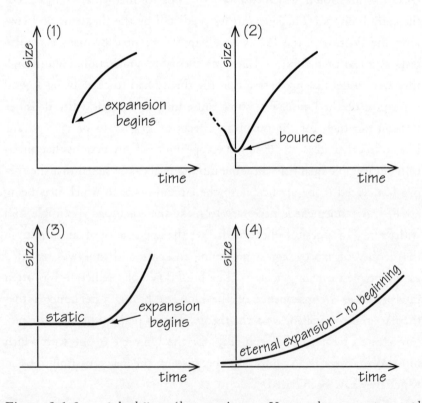

Figure 9.6 *Some of the different 'beginnings' to our Universe that are consistent with observations of its present state.*

shows some of the alternatives. They are very different conceptually and in their metaphysical ramifications, but they are all entirely compatible with all our observations of the Universe's current and past behaviour.

THE FUTURE OF THE VACUUM

"Then star nor sun shall waken,
Nor any change of light:
Nor sound of waters shaken,
Nor any sound or sight:
Nor wintry leaves nor vernal,
Nor days nor things diurnal,
Only the sleep eternal
In an eternal night."

Algernon Swinburne[42]

We have seen how the vacuum energy of the Universe may prevent the Universe from having a beginning, may influence its early inflationary moments and may be driving its expansion today, but its most dramatic effect is still to come: its domination of the Universe's future. The vacuum energy that manifests itself as Einstein's lambda force stays constant whilst every other contribution to the density of matter in the Universe – stars, planets, radiation, black holes – is diluted away by the expansion. If the vacuum lambda force has recently started accelerating the expansion of the Universe, as observations imply, then its domination will grow overwhelming in the future. The Universe will continue expanding and accelerating for ever. The temperature will fall faster, the stars will exhaust their reserves of nuclear fuel and implode to form dense dead relics of closely packed cold atoms and concentrated neutrons, or large black holes. Even the giant galaxies and clusters of galaxies will eventually follow suit, spiralling inwards upon themselves as the motions of their constituent stars are gradually slowed by the outward flow of gravitational waves and radiation. All their stars will be swallowed up in great central black holes, growing bigger until they have consumed all the material within reach. Ultimately, all these black holes will evaporate away by the Hawking evaporation process, producing a universe that contains a sea of non-interacting, fairly structureless collections of stable elementary particles and radiation. Or perhaps they do not evaporate completely, but leave a tiny relic of stable matter, or something more exotic, like a wormhole connection into another universe (or another part of our own Universe), or even a true singularity. Nobody knows.

The most fascinating thing about the cosmic vacuum energy is that, ultimately, it wins out over all other forms of matter and energy in the struggle to determine the shape of space and the rate of expansion of the universe. No matter what the structure of the universe in its earlier days before the vacuum energy comes to dominate, just as all ancient roads led to Rome so all ever-expanding universes approach a very particular accelerating universe, called the de Sitter universe after Willem de Sitter, the famous Dutch astronomer who discovered it was a solution of Einstein's theory of general relativity in 1917. It is distinguished by being the most symmetrical possible universe.

This property of an accelerating universe, that it loses all memory of how it began, is sometimes called the 'cosmic no hair property'. This curious terminology is chosen to capture the fact that all the accelerating universes become the same: they retain no individual distinguishing features (hairstyles, metaphorically speaking). This inexorable slide towards the same future state signals that there is a loss of information taking place when the universe starts accelerating. The expansion is so fast that the information content of signals sent across the universe gets degraded as fast as possible. Everything looks smoother and smoother; all differences in the rate of expansion from one direction to another are expunged at a rapid rate; no new condensations of matter can appear out of the cosmic matter distribution; local gravitational pull has lost the last battle with the overwhelming repulsion of the lambda force.

This has important consequences for any consideration of 'life' in the far future. If life requires information storage and processing to take place in some way, then we can ask whether the Universe will always permit these things to occur. When the vacuum energy is not present, and so the expansion does not ultimately accelerate, Freeman Dyson,[43] Frank Tipler and I[44] showed that there are a range of possibilities open for this rather basic form of 'life' to perpetuate itself. It can store information in elementary-particle states that are vastly better information storage repositories than those used for storing data in our present computers. In order to continue to process information indefinitely, living systems need to create and sustain deviations from perfect uniformity in the temperature and energy of the Universe.[45] This may always be possible when the accelerating vacuum energy is not present. Tiny deviations in the way in which the Universe is expanding from one direction to another can be exploited to make radiation cool at slightly different rates in different directions. The gradient in temperature thereby created can then be used to do work or process information. This does not, of course, mean that life in any shape or form *will* survive[46] for ever, let alone that it *must* survive for ever, merely that it is logically and physically possible given the known laws of physics in the absence of a vacuum energy permeating the Universe.

However, as Frank Tipler and I also showed,[47] if the vacuum energy exists then everything changes – for the worse. All evolution heads inevitably for a state of uniformity characterised by the accelerating universe of de Sitter. Information processing cannot continue for ever: it must die out. There will be less and less utilisable energy available as the material Universe is driven closer and closer to a state of uniformity. If the vacuum energy exists but there is insufficient matter in the Universe to reverse its expansion into contraction before the vacuum energy gets a grip on the expansion[48] and begins to accelerate it, then the Universe seems destined for a lifeless far future. Eventually, the acceleration leads to the appearance of communication barriers. We will be unable to receive signals from sufficiently remote parts of the Universe. It will be as if we are living inside a black hole. The part of the Universe that can affect us (or our descendants) and with which they may be in contact will be finite. In order to escape this claustrophobic future we would need the ubiquitous vacuum energy to decay. We think it must stay constant for ever, but maybe it is slowly, imperceptibly eroding. Maybe one day it will decay suddenly into radiation and ordinary forms of matter and the Universe will be left to pick up the pieces, and slowly use gravity to aggregate matter and process information. But the decay may not be so benign. We have seen that it could herald a slump into an even lower energy state for the Universe with a sudden change in the nature of physics accompanying it. It is even possible for the vacuum to decay into a new type of matter that is even more gravitationally repulsive than the lambda force. If its pressure is even more negative then something very dramatic can lie in the future. The expansion can run into a singularity of infinite density at a finite time in the future.[49]

There is one last line of speculation that must not be forgotten. In science we are used to neglecting things that have a very low probability of occurring even though they are possible in principle. For example, it is permitted by the laws of physics that my desk rise up and float in the air. All that is required is that all the molecules 'happen' to move upwards at the same moment in the course of their random movements. This is so unlikely to occur, even over the fifteen-billion-year history of the Universe, that we can forget about it for all practical purposes. However, when we have an infinite future to worry about all this, fantastically improbable physical occurrences will eventually have a significant chance of occurring. An energy field sitting at the bottom of its vacuum landscape will eventually take the fantastically unlikely step of jumping right back up to the top of the hill. An inflationary universe could begin all over again for us. Yet more improbably, our entire Universe will have some minutely small probability of undergoing a quantum-transition into another type of universe. Any inhabitants of universes undergoing such radical reform will not survive. Indeed, the probability of something dramatic of a quantum-transforming nature occurring to a system gets smaller as the system gets bigger. It is much more likely that objects within the Universe, like rocks, black holes or people, will undergo such a remake before it happens to the Universe as a whole. This possibility is important, not so much because we can say what might happen when there is an infinite time in which it can happen, but because we can't. When there is an infinite time to wait then *anything* that can happen, eventually *will* happen. Worse (or better) than that, it will happen infinitely often.

Globally, the Universe may be self-reproducing but that will merely provide other expanding regions with new beginnings. Perhaps some of their inhabitants will master the techniques needed to initiate these local inflations to order and engineer their outcomes in life-enhancing ways. For us, there is a strange symmetry to existence. The Universe may once have appeared out of the quantum vacuum, retaining a little memory of its energy. Then in the far future that vacuum energy will reassert its presence and accelerate the expansion again, this time perhaps for ever. Globally, the self-reproduction may inspire new beginnings, new physics, new dimensions, but, along our world line, in our part of the Universe, there will ultimately be sameness, starless and lifeless, for ever, it seems. Perhaps it's good that we won't be there after all.

Notes

"I must say that I find television very educational.
Whenever somebody turns it on, I go to the library
and read a book."

Groucho Marx

"There are scholars who footnote compulsively, six to a
page, writing what amounts to two books at once. There
are scholars whose frigid texts need some of the warmth
and jollity they reserve for their footnotes and other scholars who write stale, dull footnotes like the stories brought
inevitably to the minds of after-dinner speakers. There are
scholars who write weasel footnotes, footnotes that alter
the assertions in their texts. There are scholars who write
feckless, irrelevant footnotes that leave their readers dumb-
struck with confusion."

M.-C. van Leunen
A Handbook for Scholars

chapter nought

Nothingology – Flying to Nowhere

1. First words spoken in a 'talking film', *The Jazz Singer*, 1927.
2. P.L. Heath, in *The Encyclopedia of Philosophy* entry on *Nothing*, vols 5 & 6, Macmillan, NY (1967), ed. P. Edwards, p. 524.
3. Indeed, who else?

4. This word for nothing, still common in dialects in the north of England, has a Scandinavian origin in Old Norse. It had many other meanings – a bullock or an ox (the nowt-geld was a rent or tax in the north of England, payable in cattle), or a stupid or oafish person.

5. Having no proper form.

6. Worlds devoid of any unifying pattern, plan or purpose. This usage, as a contrast to the structure of a Universe, can be found in William James' *Mind*, p. 192, where he writes that 'The World . . . is pure incoherence, a chaos, a nulliverse, to whose haphazard sway . . . I will not truckle.'

7. Who reject all religious beliefs and moral precepts; sometimes expressed as an extreme form of scepticism in which all existence is denied.

8. Who maintained the heretical doctrine that in Christ's nature there was no human element, only divine nature.

9. Those who deal with things of no importance.

10. Those who do nothing.

11. Those who hold no religious or political beliefs.

12. Those who have no faith in any religious belief.

13. Those who believe that no spiritual beings exist.

14. 'Zeros' are zero dividend preference shares. These are a relatively low-risk investment issued by split capital investment trusts: low risk because they are usually first in line to be paid out when the trust is wound up. They do not pay income (hence the 'zero') but are brought to mature in different years with the gains reinvested in further zeros.

15. See, for example, the *OED*.

16. Pogo, 20 March 1965, cited by Robert M. Adams, *Nil: Episodes in the literary conquest of void during the nineteenth century*, Oxford University Press, NY (1966).

17. J.K. Galbraith, *Money: Whence it Came, Where it Went*, A. Deutsch, London (1975), p. 157.

18. *prope nihil*. There has been a persistent notion that Nothing is a very small positive quantity – much less than one expected, but still a

minimal thing. French *rien* (= nothing) is said to be derived from the Latin *rem*, the accusative singular form of the word for 'thing', and indeed Old French frequently uses *rien* in a positive sense: '*Justice amez so tôte rien*', 'Love justice above every other thing'.

19.   R.M. Adams, *Nil: Episodes in the literary conquest of void during the nineteenth century*, Oxford University Press, NY (1966), p. 3 and p. 34.

20.   R.F. Colin, 'Fakes and Frauds in the Art World', *Art in America* (April 1963).

21.   H. Kramer, *The Nation* (22 June 1963).

22.   B. Rose (ed.), *Art as Art: The Selected Writings of Ad Reinhardt*, Viking, NY (1975).

23.   J. Johns, *The Number Zero* (1959), private collection.

24.   −273.15 degrees centigrade.

25.   M. Gardner, *Mathematical Magic Show*, Penguin, London (1985), p. 24.

26.   Quoted in N. Annan, *The Dons*, HarperCollins, London (1999), p. 264.

27.   E. Maor, *To Infinity and Beyond: a cultural history of the infinite*, Princeton University Press (1987).

chapter one
Zero – The Whole Story

1.   A. Renyi, *Dialogues in Mathematics*, Holden Day, San Francisco (1976). This quote is taken from part of an imaginary Socratic dialogue.

2.   J. Boswell, *The Life of Johnson*, vol. III.

3.   R.K. Logan, *The Alphabet Effect*, St Martin's Press, NY (1986), p. 152, partly paraphrasing Constance Reid.

4.   A handy rule of thumb is Moore's Law, named after Gordon Moore, the founder of Intel, who proposed in 1965 that each new chip contained roughly twice the capacity of its predecessor and was released within 18–24 months of it.

5.   The fact that you are reading this book shows that my computers survived. But I remain unconvinced that all was down to the pre-

science of computer scientists because it was disconcerting to discover that the computers that I was told needed adjustment, did not, while those that I was assured required none, did.

6.   J.D. Barrow, *Pi in the Sky*, Oxford University Press (1992).

7.   The word 'score' has an interesting number of meanings. It refers to counting, as in keeping the score; it also refers to making a mark; and it means twenty. The score was originally the mark for this quantity on the tally stick used by the treasury.

8.   The term hieroglyph used to describe the language and number-signs of the ancient Egyptians was introduced by the Greeks. Because they could not read the symbols they found on Egyptian tombs and monuments, they believed them to be sacred signs and called them *grammata hierogluphika*, or 'carved sacred signs', hence our 'hieroglyphs'.

9.   G. Ifrah, *The Universal History of Numbers*, Harvill Press, London (1998). This is an extension and retranslation from the French original of the author's earlier volume *From One to Infinity: a universal history of numbers*, Penguin, NY (1987).

10.   The Book of Daniel 5, v. 5, 25–8. Daniel's interpretation of the writing on Belshazzar's wall.

11.   There have been many attempts to explain the existence of this sexagesimal base. It is clear that 60 is very useful for any commercial accounting involving weights, measures and fractions because it has so many factors. It may be that the arithmetic base derived from some pre-existing system of weights and measures adopted for these reasons. Another interesting possibility is that it emerged from the synthesis of two systems used by two earlier civilisations. A merger with a natural base-10 system seems unlikely since it needs a base-6 system to exist. None is known ever to have existed on Earth and there is no good reason why it should. More promising seems to be the idea that there was a merger of a base-12 and a base-5 system. Base-5 systems are natural consequences of finger-counting whilst base-12 systems are attractive for trading purposes because 2, 3, 4 and 6 are divisors of 12. We can still see the relics of its appeal in

the imperial units (12 inches in a foot, 12 old pence in a shilling, buying eggs by the dozen) and the Sumerians used it extensively in their measurements of time, length, area and volume. The words for 'one', 'two' and 'three' just correspond to the concepts one, one plus one and many: a common traditional form. But the words for 'six', 'seven', 'nine', and so on have the form of words 'five and one', 'five and two' and 'five and four'. This is evidence of a base-5 system in the past. Of course, these attractive scenarios may be entirely post hoc. The choice of 60 as the base could have been made because some autocrat had a dream, an astronomical coincidence, or a lost mystical belief in the sacredness of the number itself. We know, for instance, that some of the Babylonian gods were represented by numerals. Anu, the god of the heavens, is given the principal number, 60, because it was seen as the number of perfection. The lesser gods have other, smaller, numbers attributed to them, each has some significance. It is hard to tell whether the theological significance preceded the numerical.

12. By 2700 BC they had rotated the symbols by ninety degrees. This appears to have been because scribes had evolved from working with small hand-held tablets to large heavy slabs that could not be orientated easily in the hand, see C. Higounet, *L'Écriture*, Presses Universitaires de France, Paris (1969).

13. A further economy measure appeared about 2500 BC when very large numbers were written using a shorthand multiplication principle. For example, a number like 4×3600 would be written by putting four 'ten' markers inside the wing symbol and placing it to the right of 3600.

14. From the Latin word, *cuneus*, for 'a wedge'.

15. R.K. Guy, 'The Strong Law of Small Numbers', *American Mathematical Monthly* 95, pp. 687–712 (1988).

16. In fact, the wedge symbols were also used to extend the Babylonian system down to fractions in a mirror image of large numbers so a vertical wedge not only stood for 1, 60, 3600, etc., but also for $\frac{1}{60}$, $\frac{1}{3600}$, etc. In practice, the whole numbers were distinguished from the

fractions by writing the whole numbers from right to left in ascending size and the fractions from left to right in descending size.

17.   Non-astronomers appeared not to have done this and this misled some historians to conclude that the Babylonian zero was never used at the end of a symbol string (unlike our own zero). It was also used in the first position when employed for angular measure, so the [0;1] would denote zero degrees plus ⅟₆₀th of a degree, i.e. one minute of arc. For the most detailed analysis of the different accounting systems practised in different spheres of life in Babylonia see the detailed study by H.J. Nissen, P. Damerow, and R. Englund, *Archaic Bookkeeping: writing and techniques of economic administration in the ancient near east*, University of Chicago Press (1993).

18.   Some Babylonian texts contain subtle numerical puns, cryptograms and pieces of numerology.

19.   John Cage, 'Lecture on Nothing', *Silence* (1961).

20.   There were some differences in their system for counting time.

21.   The Mayan word for 'day' was *kin*.

22.   F. Peterson, *Ancient Mexico*, Capricorn, NY (1962).

23.   G. Ifrah, *From One to Zero*, Viking, NY (1985).

24.   B. Datta and A.N. Singh, *History of Hindu Mathematics*, Asia Publishing, Bombay (1983).

25.   See Datta and Singh, op. cit.

26.   Subandhu, quoted in G. Flegg, *Numbers Through the Ages*, Macmillan, London (1989), p. 111.

27.   The *Satsai* collection; see Datta and Singh, op. cit., p. 220.

28.   Black for an unmarried woman but indelible red for a married woman. These marks symbolise the third eye of Shiva, that of knowledge.

29.   S.C. Kak, 'The Sign for Zero', *Mankind Quarterly*, 30, pp. 199–204 (1990).

30.   Ifrah, op. cit., p. 438. These synonyms are not confined to the number zero. The Sanskrit language is rich in synonyms and all the Indian numerals possess a collection of number words taking on different images. For example, the number 2 is described by words

with meanings that span twins, couples, eyes, arms, ankles and wings.

31.   The use of zero as a number in India is displayed in both number words and number symbols. The number words were based on a decimal system and read like our own reference to 121 as 'one-two-one'. If a zero is used in this scheme it is referred to as *sunya, kha, akâsha*, or by one of the other synonyms. See A.K. Bag, *Mathematics in Ancient and Medieval India*, Chaukhambha Orientalia, Delhi (1979) and 'Symbol for Zero in Mathematical Notation in India', *Boletin de la Academia Nacional de Ciencias*, 48, pp. 254–74 (1970).

32.   J.D. Barrow, *Pi in the Sky*, Oxford University Press (1992), pp. 73–78; N.J. Bolton and D.N. Macleod, 'The Geometry of Sriyantra', *Religion* 7, pp. 66 (1977); A.P. Kulaicher, 'Sriyantra and its Mathematical Properties', *Indian Journal of History of Science*, 19, p. 279 (1984).

33.   G. Leibniz, quoted in D. Guedj, *Numbers: The Universal Language*, Thames and Hudson, London (1998), p. 59. Leibniz is credited with inventing the representation of numbers in binary form, using 0s and 1s. He describes this discovery and presents a table of the representation of powers of 2 from 2 to 2^{14} in a letter written at the end of the seventeenth century, see L. Couturat, ed., *Opuscules et fragments inédits de Leibniz*, extraits des manuscrits de la Bibliothèque Royale de Hanovre par Louis Couturat, Alcan, Paris (1903), p. 284. There appears to have been an early Indian discovery of the binary representation, perhaps as early as the second or third century AD. It was used to classify metrical verses in Vedic poetry by Pingala, see B. van Nooten, 'Binary Numbers in Indian Antiquity', *J. Indian Studies*, 21, pp. 31–50 (1993).

34.   G. Ifrah, op. cit., pp. 508–9.

35.   Plato, *The Sophist*, Loeb Classical Library, ed. H. North Fowler, pp. 336–9.

36.   T. Dantzig, *Number: The Language of Science*, Macmillan, NY (1930), p. 26.

37.   It was my choice in an electronic poll of inventions of the millennium and also for *La Repubblica*'s choice of greatest inventions.

38.   The Hebrew title was *Sefer ha Mispar*; see M. Steinschneider, *Die Mathematik bei den Juden*, p. 68, Biblioteca Mathematika (1893); M. Silberberg, *Das Buch der Zahl, ein hebräisch-arithmetisches Werk des Rabbi Abraham Ibn Ezra*, Frankfurt am Main (1895), and D.E. Smith and J. Ginsburg, 'Rabbi Ben Ezra and the Hindu-Arabic Problem', *American Mathematical Monthly* 25, pp. 99–108 (1918).

39.   He also used the Arab word *sifra*, meaning 'empty'.

40.   The oldest known European manuscript containing the Arab numerals, the *Codex Vigilanus*, comes from Logroño in northern Spain and dates from AD 976. It contains a listing of the numerals I to 9 but not the zero. It is now in the Escorial.

41.   B.L. van der Waerden, *Science Awakening*, Oxford University Press (1961), p. 58.

42.   It is interesting that in the eighth century this use of Greek numerals was allowed even when use of the Greek language was banned.

43.   The Indian zero spread East as well as West. In the eighth century, the Chinese left a gap in their representations of numbers, just like the Babylonians. Thus the number 303 is found in word form and also written in the simple 'rod' numerals as ||| |||. A circular symbol for zero did not appear until 1247 and then we find a representation of 147,000 as | ≡ ⊤0 0 0, see D. Smith, *History of Mathematics*, Dover, New York (1958), vol. 2, p. 42.

44.   *OED*.

45.   *King Lear*, Act I, scene iv.

46.   Better known to mathematicians as Fibonacci.

47.   John of Hollywood (1256), *Algorismus*, cited in *Numbers Through the Ages*, ed. G. Flegg, Macmillan, London (1989), p. 127.

48.   Cited in *Numbers Through the Ages*, op. cit., p. 127.

49.   An interesting example of counting without counting is that of a farmer who wants to make sure he has not lost any of his sheep. If he puts down one stone in a pile when each sheep enters the field in the morning and then removes one from the pile when each sheep leaves the field at dusk then he only needs to check that there are no stones remaining when the last sheep has left. This example of how the

farmer can use mathematics without understanding it in some respects is reminiscent of John Searle's 'Chinese Room' argument against the possibility of artificial intelligence having semantic ability; see J.R. Searle, *The Mystery of Consciousness*, Granta, London (1997).

50. K. Menninger, *Number Words and Number Symbols*, MIT Press, MA (1969).

51. Until the development of the first calculating machines in England during the 1940s, the word 'computer' was solely a description of a person who performed calculations. It was then adopted to describe machines that can compute. Ironically, today it is used only to describe non-human calculation. During the interim period there have been many other mechanical 'adding machines' and non-programmable devices which became known as 'calculators'. The word computer as a description of a human calculator derives from *computare*, the medieval Latin for 'cut', which echoes the cutting of notches on a tallying stick in the way that the English word 'score' means to keep count and make a mark (and the quantity 20). The medieval Latin for 'calculate' was *calculare*, and the Latin *calculus ponere* meant to move or place pebbles, an echo of the movement of stones as counters on a counting board. We recognise the source here for the naming of the differential and integral calculus, invented by Newton and Leibniz (and also for Hergé's Professor Calculus).

52. P.A.M. Dirac, *The Principles of Quantum Mechanics*, Oxford University Press (1958).

chapter two

Much Ado About Nothing

1. Leonardo da Vinci, *The Notebook*, translated and edited by E. Macurdy, London (1954), p. 61.

2. U. Eco, *The Name of the Rose*, Secker & Warburg, London (1983).

3. Lyric from 'Me and Bobby McGee' (1969).

4. This phrase has made several appearances in recent years in popular expositions of modern physics (see for example the books by

Paul Davies and James Trefil, whose titles use it). It is used as a synonym for the laws of Nature or the underlying structural features of the Universe which may be partly (or wholly) independent of the laws, for example the existence and dimensionality of time and space.

5.   The only slightly counter-intuitive property is the requirement that we define zero factorial, $0! = 1 = 1!$ since the factorial operation is defined recursively by $(n + 1)! = (n + 1) \times n!$

6.   See J.D. Barrow, *Pi in the Sky*, Oxford University Press (1992), pp. 205–216, for a more detailed account of these manipulations.

7.   David Hilbert was one of the world's foremost mathematicians in the first part of the twentieth century.

8.   M. Friedman, ed., *Martin Buber's Life and Work: The Early Years 1878-1923*, E.P. Dutton, NY (1981).

9.   J.-P. Sartre, *Being and Nothingness* (transl. H. Barnes), Routledge, London (1998), p. 16.

10.   Sartre, op. cit., p. 15.

11.   B. Rotman, *Signifying Nothing: The Semiotics of Zero*, Stanford University Press (1993), p. 63.

12.   *The Odyssey*, Book IX, lines 360–413, *Great Books of the Western World*, vol. 4, Encyclopaedia Britannica Inc., University of Chicago (1980).

13.   In Greek οὐτις, meaning nobody.

14.   G.S. Kirk and J.E. Raven, *The Presocratic Philosophers: a critical history with a selection of texts*, Cambridge University Press (1957).

15.   Fragment quoted by S. Sambursky, *The Physical World of the Greeks*, Routledge, London (1987), pp. 19–20.

16.   Sambursky, op. cit., p. 22.

17.   Sambursky, op. cit., p. 108.

18.   B. Inwood, 'The origin of Epicurus' concept of void', *Classical Philology* 76, pp. 273–85 (1981); D. Sedley, 'Two conceptions of vacuum', *Phronesis*, 27, pp. 175–93 (1982).

19.   Except for a few fragments, Leucippus' and Democritus' writings do not survive and his ideas have been partially reconstructed from the commentary of others, particularly Lucretius, Aristotle and the lat-

ter's successor as head of the Academy, Theophrastus. When considering his views on whether atoms could be observed or not it is interesting to refer to some of the fragments of Democritus' writings that do survive. He appears to have adopted a rather 'modern' (at least nineteenth-century) Kantian view that there is a distinction between what we can know about things and their real nature: 'It will be obvious that it is impossible to understand how in reality each thing is,' he writes, for 'we know nothing accurately in reality, but as it changes according to the bodily conditions, and the constitution of things that flow upon the body and impinge upon it . . . for truth lies in an abyss.' S. Sambursky, op. cit., p. 131.

20. Democritus endowed his atoms only with the properties of size and shape; Epicurus also allowed them to have weight in order to determine aspects of their motion under gravity.

21. Lucretius II, 308–322.

22. Aristotle quoted in J. Robinson, *An Introduction to Greek Philosophy* (1968), Boston, p. 75.

23. S. Sambursky, *Physics of the Stoics*, Routledge, London (1987); R.B. Todd, 'Cleomedes and the Stoic conception of the void', *Apeiron*, 16, pp. 129–36 (1982).

24. F. Solmsen, *Aristotle's System of the Physical World*, Cornell University Press, Ithaca (1960); R. Sorabji, *Matter, Space & Motion*, Duckworth, London (1988); E. Grant, *Much Ado About Nothing: Theories of Space and Vacuum from the Middle Ages to the Scientific Revolution*, Cambridge University Press (1981).

25. Quoted by C. Pickover, *The Loom of God*, Plenum, NY (1997), p. 122.

26. In the nineteenth and twentieth centuries, mathematicians have come to appreciate the systematic way of constructing so-called 'space-filling' curves which will ultimately pass through every point in a specified region.

27. See for example *The Complete Works of John Davies of Hereford*, Edinburgh (1878) and for a review, V. Harris, *All Coherence Gone*, University of Chicago Press (1949).

28. The *efficient* cause in Aristotle's sense.

29. For a detailed discussion of these arguments see W.L. Craig, *The Cosmological Argument From Plato to Leibniz*, Macmillan, London (1980).

30. R. Adams, *Nil: Episodes in the literary conquest of void during the nineteenth century*, Oxford University Press, NY (1966), p. 33.

31. For example, on the ground that if empty space were a body then when another body were placed in empty space there would be two bodies at the same place at the same time, and if two bodies could be coincident like this then why not all bodies, which he regarded as absurd.

32. It is intriguing to note that Aristotle formulates what we now call Newton's first law of motion, that bodies acted on by no forces move at constant velocity, but rejects it as a reductio ad absurdum.

33. It was rediscovered at this time. Lucretius' *De Rerum Natura*, where it appears in Bk I, p. 385, was unknown in Europe until the fifteenth century.

34. *De Rerum Natura*, Bk I, pp. 385–97.

35. For a classic study of these and a host of other medieval investigations into the nature of space, infinity and the void, see the beautiful book by Edward Grant, *Much Ado About Nothing: Theories of Space and Vacuum from the Middle Ages to the Scientific Revolution*, Cambridge University Press (1981), p. 83.

36. Grant, op. cit., p. 89.

37. This problem is discussed by Galileo in his *Discourse Concerning Two New Sciences* on the first day.

38. Here one is reminded of the modern idea, introduced by Roger Penrose, of cosmic censorship, the idea that Nature abhors the creation of singularities in space time which are visible from far away and which can causally influence events there. The 'cosmic censor' (not a person, just an internal property of Einstein's equations, suspected to be necessary for the physical self-consistency of Einstein's theory of general relativity) is hypothesised to cloak all singularities that could form with an event horizon. This horizon prevents information from the singularity, where the laws of physics break down, from passing out to affect events far from the singularity. The sim-

plest example of this device is that of the black hole where, unless quantum gravitational effects always intervene to prevent an actual physical singularity of infinite density forming at the centre of the black hole, an event horizon always stops outside observers seeing it or being causally influenced by it.

39.   Aristotle did not believe in the existence of this imaginary extracosmic void, of course, and commented that some people wrongly deduced its existence merely because they were unable to imagine an end to some things.

40.   This quote is usually attributed to Nicholas of Cusa but was widely cited as early as the twelfth century. Grant, op. cit., pp. 346-7, gives Alan of Lille as the first known source; for a detailed study of the question see also D. Mahnker, *Unendliche Sphäre und Allmittelpunkt*, Hale/Salle: M. Niemeyer Verlag, 1037 (1937), pp. 171–6.

41.   For a detailed account of these Design Arguments, see J.D. Barrow & F.J. Tipler, *The Anthropic Cosmological Principle*, Oxford University Press (1986).

42.   I. Newton, *Opticks*, Book III Pt.I, *Great Books of the Western World*, vol. 34, W. Benton, Chicago (1980), pp. 542–3.

43.   See E. Grant, op. cit., p. 245; A. Koyré, *From the Closed World to the Infinite Universe*, p. 297 note 2, Johns Hopkins Press, Baltimore (1957); and W.G. Hiscock, ed., *David Gregory, Isaac Newton and their Circle: Extracts from David Gregory's Memoranda, 1667–1708*, Oxford (1937), printed for the editor.

44.   Quoted by Robert Lindsay on *Parkinson*, BBC Television, 15 January 1999.

45.   R.L. Colie, *Paradoxia Epidemica: the Renaissance tradition of paradox*, Princeton University Press, NJ (1966), pp. 223–4. See also A.E. Malloch, 'The Techniques and Function of the Renaissance Paradox', *SP*, 53, pp. 191–203 (1956).

46.   Attributed to Edward Dyer by most commentators and to Edward Daunce by R.B. Sargent in *The Authorship of The Prayse of Nothing, The Library*, 4th series, 12, pp. 322–31 (1932); the passage quoted here is the second stanza. See also H.K. Miller, 'The Paradoxical Encomium

with Special Reference to its Vogue in England, 1660–1800', *MP*, 53, p. 145 (1956).

47.   *Facetiae*, chap. 2, pp. 389–92, London (1817), cited by Colie, op. cit., p. 226.

48.   J. Passerat, *Nihil*, quoted by R. Colie, op. cit., p. 224.

49.   Act I, scene 2, line 292.

50.   The main sources for the story are a translation of a novella by the Italian short-story writer Matteo Bandello (1485–1561) and Ludovico Ariosto's *Orlando Furioso*, an Italian epic poem (1532).

51.   P.A. Jorgenson, 'Much Ado about Nothing', *Shakespeare Quarterly*, 5, pp. 287–95 (1954).

52.   *Much Ado About Nothing*, Act 4, scene 1, line 269.

53.   *Macbeth*, I, iii, 141–2.

54.   *Macbeth*, V, v, 16.

55.   Colie, op. cit., p. 240.

56.   *Hamlet*, III, ii, 119–28.

57.   *King Lear*, I, i, 90.

58.   R.F. Fleissner, 'The "Nothing" Element in *King Lear*', *Shakespeare Quarterly*, 13, pp. 62–71 (1962); H.S. Babb, '*King Lear*: the quality of nothing', in *Essays in Stylistic Analysis*, Harcourt, Brace, Jovanovich, NY (1972).

59.   It has also been suggested by David Willbern that in the theatre of Shakespeare's day, Nothing would have sounded like 'noting' (thus, 'Much Ado about Noting') and this would have added a further contrasting meaning, the sense of 'noting' being our usual one together with observing, eavesdropping and overhearing, see R.G. White, *The Works of William Shakespeare*, Boston (1857), III, p. 226, but this seems to sell short the ingenuity of Shakespeare's multiple meanings and creates a less enticing title. This idea does not seem to have been taken up by other commentators; see D. Willbern, 'Shakespeare's Nothing', in *Representing Shakespeare*, eds M.M. Schwarz & C. Kahn, Johns Hopkins University Press, Baltimore (1980), pp. 244–63, and B. Munari, *The Discovery of the Circle*, transl. M. & E. Maestro, G. Witterborn, NY (1966) and H. Kökeritz, *Shakespeare's Pronunciation*, Yale University Press, New Haven (1953). Willbern

also pursues the psychoanalysis of Shakespeare to what many will consider to be an unconvincing extent. There are several other studies of this general sort which have looked at aspects of Shakespeare's sense of Nothing, see D. Fraser, 'Cordelia's Nothing', *Cambridge Quarterly*, 9, pp. 1–10 (1978), L. Shengold, 'The Meaning of Nothing', *Psychoanalytic Quarterly*, 43, pp. 115–19 (1974).

60. Sartre, op. cit., p. 23. Note that 'nihilate' (*néantir*) is defined by Sartre as 'nihilation is that by which consciousness exists. To nihilate is to encase with a shell of non-being.'

61. The curtain was brought down by John Dunton's huge compendium *Athenian Sport: or, Two Thousand Paradoxes merrily argued to Amuse and Divert the Age* (1701).

62. G. Galileo, *Dialogue Concerning Two World Systems*, transl. S. Drake, California University Press, Berkeley (1953).

63. Galileo, op. cit., pp. 103–4.

64. The medieval historian Edward Grant remarks that '. . . approximately two thousand Latin manuscripts of the work of Aristotle have been identified. If this number of manuscripts survived the rigors of the centuries, it is plausible to suppose that thousands more have perished. The extant manuscripts are a good measure of the pervasive hold that the works of Aristotle had on the intellectual life of the Middle Ages and Renaissance. With the possible exception of Galen . . . , no other Greek or Islamic scientist has left a comparable manuscript legacy.' E. Grant, *The Foundations of Modern Science in the Middle Ages*, Cambridge University Press, NY (1996), pp. 26–7.

chapter   three

Constructing Nothing

1. Radio 3 broadcast *Close Encounters with Kurt Gödel*, reviewed by B. Martin, *The Mathematical Gazette* (1986), p. 53.

2. One of the curious facts about this view of the physical make-up of the world is that it arose amongst the Stoics as purely a religious belief at a time when there neither was, nor could be, any experimen-

tal evidence in its favour. However, it has turned out to be correct in its general conception of the hierarchical structure of matter.

3. G. Galileo, *Dialogues Concerning Two New Sciences* (1638), Britannica Great Books, University of Chicago (1980), p. 137. See also C. Webster, *Arch. Hist. Exact. Sci.*, 2, p. 441 (1965).

4. This is what Aristotle would have called an 'efficient cause'.

5. A British gold coin with a face value of one pound, minted first in 1663 for trade with Africa. After 1717 it became legal tender in Britain with a value fixed at twenty-one shillings or £1.05 in present UK currency. It is still used by auction houses and to fix the prize money of some horse races.

6. Letters announcing the discoveries of E. Torricelli are translated in V. Cioffair, *The Physical Treatises of Pascal etc.*, Columbia University Press, NY (1937), p. 163. The originals are in E. Torricelli, *Opera*, Faenza, Montanari (1919), vol. 3, pp. 186–201.

7. If the column of mercury has height h, density d, the acceleration due to gravity is g and the tube has cross-sectional area A, then the downward force of the weight of mercury in the column is given by $hAdg$. When the mercury column comes into equilibrium, this downward force is balanced by the upward force due to the pressure exerted by the column, P, and this equals PA. Notice that both forces are proportional to A and so the height of the mercury in equilibrium is the same regardless of the value of the area A.

8. W.E.K. Middleton, *The History of the Barometer*, Johns Hopkins Press, Baltimore (1964).

9. S. Sambursky, *Physical Thought from the Presocractics to Quantum Physics*, Hutchinson, London (1974), p. 337.

10. S.G. Brush, ed., *Kinetic Theory*, vol. 1, Pergamon, Oxford (1965), contains extracts from Boyle's original papers; see in particular 'The Spring of the Air' from his book *New Experiments Physico-Mechanical, touching the spring of the air, and its effects*. Boyle's work on air pressure is discussed in M. Boas Hall, *Robert Boyle on Natural Philosophy*, Indiana University Press, Bloomington (1965); R.E.W. Maddison, *The Life of the Honourable Robert Boyle, F.R.S.*, Taylor & Francis, London (1969);

J.B. Conant, ed., *Harvard Case Histories in Experimental Science*, Harvard University Press, Cambridge Mass. (1950). For an excellent historical overview, see S.G. Brush, *The Kind of Motion We Call Heat: a history of the kinetic theory of gases in the 19th century*, vol. I, New Holland, Amsterdam (1976).

11. M. Boas Hall, *Robert Boyle on Natural Philosophy*, Indiana University Press, Bloomington (1965).

12. This was based upon the deduction that the product of the pressure and volume occupied by the gas remains a constant when both change without altering the temperature. This result has become known as Boyle's Law although he did not discover it himself (or claim to have). He merely confirmed the earlier experiments of Richard Townley. For the story, see C. Webster, *Nature*, 197, p. 226 (1963), and *Arch. Hist. Exact. Sci.*, 2, p. 441 (1965).

13. A German translation of O. von Guericke, *Experimenta nova (ut vocantur) Magdeburgica de vacuo spatio primum a R.P. Gaspare Schotto*, Amsterdam (1672), was made by F. Danneman, *Otto von Guericke's neue 'Magdeburgische' Versuche über den leeren Raum*, Leipzig (1894).

14. O. von Guericke, *The New (so-called) Magdeburg Experiments of Otto von Guericke*, translated by M.G.F. Adams, Kluwer, Dordrecht (1994), original publication 1672 by K. Schott, Würzburg, front plate.

15. O. von Guericke, op. cit., p. 162.

16. The second volume of his Treatise expounded his opinions about the nature and extent of void space. He believed in a universe of stars surrounded by an infinite void.

17. O. von Guericke, *Experimenta nova*, p. 63. The translation is from E. Grant, *Much Ado About Nothing*, p. 216.

18. A. Krailsheimer, *Pascal*, Oxford University Press (1980), p. 18.

19. B. Pascal, *Pensées*, trans. A. Krailsheimer, Penguin, London (1966).

20. Pascal planned a book on the vacuum entitled *Traité du vide*, but it was never completed. The Preface exists but the remaining parts have been lost. Two posthumous papers appeared in 1663, one on the subject of barometric pressure, *L'Équilibre des liqueurs*, the other about the hydraulic press, entitled *La Pesanteur de la masse d'air*.

21. Blaise Pascal, by Philippe de Champagne, engraved by H. Meyer; reproduced by permission of Mary Evans, Picture Library.

22. Spiers, I.H.B. & A.G.H. (transl.), *The Physical Treatise of Pascal*, Columbia University Press, NY (1937), p. 101.

23. Adapted from H. Genz, *Nothingness*, Perseus Books, Reading, MA (1999), p. 113.

24. *Independent*, 15 April 2000.

25. Second letter of Noël to Pascal, in B. Pascal, *Oeuvres*, eds I. Brunschvicq and P. Boutroux, Paris (1908), 2, pp. 108–9, transl. R. Colie, *Paradoxia Epidemica*, Princeton University Press (1966), p. 256.

26. *Oeuvres*, 2, pp. 110–11.

27. This is because the Universe is expanding, hence its size is linked to its age. In order for nuclei of elements heavier than hydrogen and helium to have sufficient time to form in stars, billions of years are needed and so the Universe must be billions of light years in size, see J.D. Barrow & F.J. Tipler, *The Anthropic Cosmological Principle*, Oxford University Press (1986).

28. G. Stein, *The Geographical History of America* (1936).

29. F. Hoyle, *Observer*, 9 September 1979.

30. Defined by the distance that light has been able to travel during the age of the Universe, since its expansion began, about 13 billion years.

31. It is possible for the dark matter to be supplied by much lighter neutrinos which we already know to exist. We have only upper limits on their possible masses. These experimental limits are very weak. However, although these light neutrinos could supply the quantity of dark matter required in a natural way, they cause the luminous matter to cluster into patterns that do not look like those displayed by populations of real galaxies. Large computer simulations show that, in contrast, the much heavier neutrino-like particles (WIMPS = weakly interacting massive particles) seem to produce a close match to the observed clustering of luminous galaxies.

chapter four

The Drift Towards the Ether

1. D. Gjertsen, *The Newton Handbook*, Routledge & Kegan Paul, London (1986), p. 160.

2. Newton's 1st law does not hold for observers who are in a state of accelerated motion relative to 'absolute space'; for example, if you look out of the window of a spinning rocket you will see objects rotating about you, and hence apparently accelerating, even though they are acted upon by no forces. Thus Newton's laws will be seen to be true only for a special class of cosmic observers, called 'inertial' observers, who are moving so that they are not accelerating relative to 'absolute space'. One of the ways in which Einstein's general theory of relativity supersedes Newton's is that it provides laws of gravity and of motion which are true for all observers regardless of their motion: there are no observers for whom the laws of Nature are always simpler than they appear for others. See J.D. Barrow, *The Universe that Discovered Itself*, Oxford University Press (2000), pp. 108–24, for a fuller discussion of this development.

3. Experimental accuracy did not permit the detection of the very small change in the fall of a body in air compared to that in a vacuum.

4. *Opticks* (1979 edn), p. 349.

5. Bentley, a distinguished classical scholar, sought Newton's advice when preparing his Boyle Lectures on natural theology. He was anxious to propose a new form of the argument from design, in which he would claim that it was the special mathematical forms of the laws of motion and gravity that were evidence for the existence of an intelligent Designer – a view with which Newton did not disagree. For a detailed discussion of this and other arguments of this sort, see J.D. Barrow & F.J. Tipler, *The Anthropic Cosmological Principle*, Oxford University Press (1986).

6. I.B. Cohen, *Isaac Newton's Papers and Letters on Natural Philosophy*, Harvard University Press (1958), p. 279, letter of 25.2.1693.

7.   R. Descartes, *The World, or a Treatise on Light* (1636).

8.   Much has been made of the role that 'beauty' or some other human opinion of 'elegance' or 'economy' plays in the physicist's conception of Nature (see, for example, S. Chandrasekhar, 'Beauty and the Quest for Beauty in Science', *Physics Today*, July 1979, pp. 25–30); however, this is often over-romanticised by physicists long after the creative process took place. Freeman Dyson has an interesting opinion of the work of Dirac and Einstein in this respect, arguing that their most important work was not guided by aesthetic considerations, but by experiment. Moreover, when they did become overtaken by the quest of beauty in their equations their useful scientific contributions ceased. Another interesting remark on the aesthetic appeal of Einstein's theory was made by the experimental physicist, and operationalist philosopher, Percy Bridgman in his book *Reflections of a Physicist*, Philosophical Library Inc., New York (1950). He regarded the search for 'beautiful' equations to be a dangerous metaphysical diversion: 'The metaphysical element I feel to be active in the attitude of many cosmologists to mathematics. By metaphysical I mean the assumption of the "existence" of validities for which there can be no operational control . . . At any rate, I should call metaphysical the conviction that the universe is run on exact mathematical principles, and its corollary that it is possible for human beings by a fortunate *tour de force* to formulate these principles. I believe that this attitude is back of the sentiment of many cosmologists towards Einstein's differential equations of generalised relativity theory – when, for example, I ask an eminent cosmologist in conversation why he does not give up the Einstein equations if they make him so much trouble, and he replies that such a thing is unthinkable, that these are the only things that we are really sure of.'

9.   *Opticks*, Query 18.

10. Op. cit., Query 21.

11. As he had first proposed to Boyle many years earlier.

12. *Opticks*, Query 21.

13. Remark to George FitzGerald of Trinity College Dublin, 1896.

14.   We call this Olbers' Paradox although Edmund Halley (famous for discovering the periodicity of the comet that now bears his name) appears to have been the first astronomer to highlight its significance, calling it a 'metaphysical paradox', in 1714. For an illuminating discussion of the dark sky paradox, see E.R. Harrison, *Darkness at Night*, Harvard University Press (1987). The account in S. Jaki, *The Paradox of Olbers' Paradox*, Herder and Herder, NY (1969), is not recommended and resolution to the paradox suggested therein is incorrect, see Harrison, op.cit., p. 173.

15.   E.R. Harrison, *Darkness at Night*, Harvard University Press (1987), p. 69.

16.   J.E. Gore, *Planetary and Stellar Studies*, Roper and Drowley, London (1888).

17.   J.E. Gore, op.cit., p. 233, cited by E.R. Harrison, *Darkness at Night*, pp. 167–8.

18.   S. Newcomb, *Popular Astronomy*, Harper, NY (1878).

19.   Figure adapted from E.R. Harrison, *Darkness at Night*, p. 169.

20.   The *Independent* newspaper, Saturday magazine supplement, 17 January 1998, p. 10.

21.   This version of the Design Argument made its first considered appearance in Bentley's Boyle Lectures. These lectures were very significantly informed by the letters from Newton to Bentley. Newton was extremely sympathetic to his work being used for such religious apologetics even though he did not publish on this subject himself; for a detailed discussion see J.D. Barrow & F.J. Tipler, *The Anthropic Cosmological Principle*, Oxford University Press (1986).

22.   J. Cook, *Clavis naturae; or, the mystery of philosophy unvail'd*, London (1733), pp. 284–6.

23.   W. Whewell, *Astronomy and General Physics considered with reference to natural theology*, London (1833). The 3rd Bridgewater Treatise.

24.   Whewell argued in some detail for the appeal of the ether hypothesis by pointing to the simplicity of the hypothesis of an ether governed by mechanical laws when compared with the complexity of the optical phenomena that it was able to explain. Elsewhere,

Whewell supposes that there must exist several different ethereal fluids in order to explain the different propagation properties of sound, electricity, magnetism and chemical phenomena because their effects appear to be so qualitatively different.

25. B. Stewart and P.G. Tait, *The Unseen universe; or, physical speculations on a future state*, London (1875).

26. F. Kafka, *Parables*.

27. Fresnel's ether was stationary. Stokes imagined that the Earth dragged the ether along as it rotated on its axis each day and orbited the Sun annually. Maxwell proposed an ether that was a magneto-electric medium consisting of a fluid filled with spinning vortex tubes as a model of the electromagnetic field.

28. B. Jaffe, *Michelson and the Speed of Light*, Doubleday, NY (1960); H.B. Lemon and A.A. Michelson, *The American Physics Teacher*, 4, pp. 1–11, Feb. (1936); R.A. Millikan and A.A. Michelson, *The Scientific Monthly*, 48, pp. 16–27, Jan. (1939).

29. J.R. Smithson, 'Michelson at Annapolis', *American Journal of Physics* 18, 425–8 (1950).

30. J.C. Maxwell, *Encyclopaedia Britannica* (9th edn), article on 'The Ether'.

31. The interference of light was first demonstrated by Thomas Young in 1803.

32. We are assuming that one of the light paths is aligned with the direction of motion of the ether. In general it would not be, but this is easily incorporated into the calculation and does not alter the significance of a null result.

33. A. Michelson, 'The Relative Motion of the Earth and the Luminiferous Ether', *American Journal of Science*, series 3, 22, pp. 120–9 (1881).

34. R.S. Shankland, 'Michelson at Case', *American Journal of Physics* 17, pp. 487–90 (1950).

35. A. Michelson and E. Morley, *American Journal of Science*, series 3, 34, pp. 333–45 (1887).

36. Lorentz proposed that the values of mass and time are also changed. The transformation for mass, length and time are now generally

known as the Lorentz transformations and form part of Einstein's special theory of relativity.

37. Lorentz seems to have regarded the ether as inadequate as a representation of the vacuum. It had too many attributes; see A.J. Knox, 'Hendrik Antoon Lorentz, the Ether, and the General Theory of Relativity', *Archive for History of Exact Sciences*, 38, pp. 67–78 (1988).

38. Radio conversation released by UK Chief of Naval Operations, quoted in *The Bilge Pump*, the newsletter of the Sunshine Coast Squadron in British Columbia, October 1994.

39. A. Einstein, 'Zur Elektrodynamick bewegter Körper', *Annalen der Physik*, 17, pp. 891–921. Librarians often remove this volume from open library shelves because of the risk of theft.

40. M for matrix or mystery.

41. This discussion should be compared with that of 'paradigms' introduced by the late Thomas Kuhn, and which is popular in some circles. Kuhn made popular the idea of scientific 'revolutions' in which new paradigms periodically sweep away old ones. Kuhn's thinking was strongly influenced by his historical studies of the Copernican 'revolution' which overthrew the Ptolemaic system of astronomy that preceded it. However, this example was special and not typical of the evolution of theories of physics which we see from Newton onwards. The evolution of those theories did not involve the overthrow of the old theory or paradigm. Rather the old theory was revealed to be a limiting case of the new, more general, more widely applicable, theory.

42. Cited in Jaffe, op.cit, p. 168.

43. A. Einstein, 'Über die Untersuchung des Ätherzustandes im magnetischen Felde', *Physikalische Blätter*, 27, pp. 390–1 (1971).

44. Einstein Archive FK 53, Letter to M. Maric, July 1899.

45. Einstein denied the existence of a physical ether consistently between 1905 and 1916 in scientific articles and in the popular press.

chapter five

Whatever Happened to Zero?

1.   A. Marvell, *The Poetical Works of Andrew Marvell*, Alexander Murray, London (1870), 'Definition of Love', stanza VII.

2.   The philosopher Immanuel Kant argued that Euclidean geometry was the only geometry that is humanly thinkable. It was forced upon us like a straitjacket by the way minds work. This was soon shown to be totally incorrect by the creation of new geometries. In fact, Kant should not have needed new mathematical developments to tell him this. By looking at any Euclidean geometrical example (for example a triangle on a flat surface) in a curved mirror it should have been clear that the laws of reflection guarantee that there must exist geometrical 'laws' on the curved surface which are reflections of those that exist on the flat surface.

3.   Euclid, *Elements*, *Great Books of the Western World*, Encyclopaedia Britannica Inc., Chicago (1980), vol. 11.

4.   Euclid's original axiom stated that 'If a line A crossing two lines B and C makes the sum of the interior angles on one side of A less than two right angles, then B and C meet on that side.' A simpler statement found in many geometry textbooks has the form 'through any point not on a given line L there passes exactly one line parallel to L'. Euclid's other four postulates were that: 1. It is possible to draw a straight line from any point; 2. It is possible to produce a finite straight line continuously in a straight line; 3. It is possible to describe a circle with any centre and radius; 4. All right angles are equal to one another.

5.   B. de Spinoza, *Ethics* (1670), in *Great Books of the Western World*, vol. 31, Encyclopaedia Britannica Inc, Chicago (1980).

6.   This can be done either by stating that through any point not on a given line L there must pass more than one line parallel to L, or no lines parallel to L.

7.   If it is possible to deduce that $0 = 1$, the system is inconsistent. Note that if any false statement of this sort is derivable then one can use it to deduce that *any* statement holds in the language of the system.

8.   J. Richards, 'The reception of a mathematical theory: non-Euclidean geometry in England 1868–1883', in *Natural Order: Historical Studies of Scientific Culture*, eds B. Barnes and S. Shapin, Sage Publications, Beverly Hills (1979); E.A. Purcell, *The Crisis of Democratic Theory*, University of Kentucky Press, Lexington (1973); J.D. Barrow, *Pi in the Sky*, Oxford University Press (1992).

9.   R.L. Graham, D.E. Knuth & O. Patashnik, *Concrete Mathematics*, Addison Wesley, Reading (1989), p. 56.

10.   In fact, Euclid's intuitively selected axioms were found to contain some strange omissions. For example, only in 1882 did Moritz Pasch notice that some things that seemed 'obviously' true could not be proved from Euclid's classical axioms. One example is the following: if A, B, C and D are points arranged on a line so that B lies between A and C, and C lies between B and D, then show that B has to lie between A and D. This has to be added to Euclidean geometry as an additional axiom, if it is needed. Other observed facts which Euclid did not formulate as axioms, but which cannot be established from his chosen axioms, are that an unending straight line passing through the centre of a circle must intersect the perimeter of the circle and that a straight line that intersects one side of a triangle, but which does not intersect any of the triangle's vertices, must intersect one of its other sides.

11.   One of the strange things about the discovery of non-Euclidean geometries by mathematicians is that it took so long and proved so controversial. Artists and sculptors had discovered the rules that govern lines and angles on curved surfaces centuries earlier. In my book *Pi in the Sky*, there is a picture of an early Indian meditation symbol, a Sri Yantra, in which a nested pattern of triangles is arranged so that many lines intersect at a single point. Such objects were commonly drawn on flat surfaces but this one, made of rock salt, is unusual in that it was made on a curved spherical surface and must have required considerable appreciation of non-Euclidean geometry in order to be made. Another interesting factor that is hard to reconcile with the slowness of mathematicians to catch on to

non-Euclidean geometries is the existence of curved mirrors, of glass or of polished metal. If you look in a curved mirror at a right-angled Euclidean triangle drawn on a flat surface then you are seeing a direct mapping of that triangle and the rules that govern its properties (like Pythagoras' theorem) on to a curved surface. The rules themselves have direct reflected counterparts in the distorted triangle that is seen in the mirror. This tells you that there must be a set of rules governing the properties of the triangle in the mirror. This interrelationship was eventually captured more formally by Beltrami, Poincaré and Klein, who showed that Euclidean and non-Euclidean geometries are equiconsistent, that is, the logical self-consistency of one demands the self-consistency of the other.

12. It is easy to see that this element must be unique. For if there were two elements, I and J, with the identity property then I must equal I combined with J which must equal J, so I is the same as J.

13. If we included zero then we could form fractions like $\frac{x}{0}$ which are not finite fractions and closure would be violated.

14. In fact, the German mathematician Felix Klein initiated a programme in 1872 (the so-called 'Erlangen programme') which aimed to unify the study of all geometries by defining them as mathematical structures with certain transformation properties. For example, one might define Euclidean geometry as the study whose properties remain the same under rotations, reflections, similarities and translations in space.

15. Unexpectedly, the Austrian mathematician Kurt Gödel showed that if a mathematical structure is rich enough to contain arithmetic then it is not possible that its defining axioms are inconsistent. If they are assumed to be consistent then the structure is necessarily incomplete in the sense that there must exist statements framed in the language of the structure which can neither be proved to be true nor false using the rules of reasoning of the system. Euclidean and non-Euclidean geometries are not rich enough to contain the structure of arithmetic and so this incompleteness theorem does not apply to them; see J.D. Barrow, *Impossibility: the limits of science and the sci-*

ence of limits, Oxford University Press (1998), chapter 8, for more details.

16. This freedom to specify axioms allows a statement to be 'true' in one axiomatic system but 'false' in another.

17. It should be noted that, although the simple mathematical structure of a group that we have introduced requires the existence of an identity element which looks like the zero or arithmetic in some cases, not all mathematical structures have a zero element.

18. F. Harary & R. Read, *Proc. Graphs and Combinatorics Conference,* George Washington University, Springer, NY (1973).

19. M. Gardner, *Mathematical Magic Show,* Penguin, London (1977).

20. B. Reznick, 'A Set is a Set', *Mathematics Magazine,* 66, p. 95 (April issue 1993).

21. The full title was *An Investigation of the Laws of Thought on Which are Founded the Mathematical Theories of Logic and Probabilities.* He also developed some of these ideas in his earlier book *The Mathematical Analysis of Logic.*

22. This is obviously the case for finite sets and (not so obviously) is also the case for infinite sets as well, as proved by Georg Cantor. It means that there is a never-ending ascending staircase of infinities, each infinitely bigger (in the well-defined sense of there not being a one-to-one correspondence between the members) than the previous one. The set of all subsets of a given set is called its *power set.*

23. These diagrams are named after their inventor, John Venn (1834–83).

24. The basic idea of this construction was discovered by the German logician Gottlob Frege and then rediscovered by Bertrand Russell. The form presented here is simpler in its treatment and was introduced as a refinement of Frege's scheme by John von Neumann.

25. R. Cleveland, 'The Axioms of Set Theory', *Mathematics Magazine,* 52, 4, pp. 256–7 (1979).

26. R. Rucker, *Infinity and the Mind,* Paladin, London (1982), p. 40.

27. D.E. Knuth, *Surreal Numbers: How Two Ex-Students Turned On to Pure Mathematics and Found Total Happiness,* Addison Wesley, NY (1974). In this quote you notice that Conway's initials conveniently provide the Hebrew consonants for Jehovah = Yahweh.

28.   J.H. Conway, *On Numbers and Games*, Academic, NY (1976).

29.   D.E. Knuth, op.cit.

30.   In the postscript to the book (p. 113) Knuth writes, 'I decided that creativity can't be taught using a textbook, but that an "anti-text" such as this novel might be useful. I therefore tried to write the exact opposite to Landau's *Grundlagen der Mathematik*; my aim was to show how mathematics can be "taken out of the classroom and into life", and to urge the reader to try his or her own hand at exploring abstract mathematical ideas.' Knuth picks on Landau during the dialogues but his most general target is probably the Bourbaki approach to presenting mathematics.

31.   Negative numbers are defined analogously $-x = \{-R \mid -L\}$.

32.   If x and y are given by $x = \{x^L \mid x^R\}$ and $y = \{y^L \mid y^R\}$ then the sum $x + y = \{x^L + y, x + y^L \mid x^R + y, x + y^R\}$ and the product $xy = \{x^L y + x\, y^L - x^L\, y^L, x^R\, y + x\, y^R - x^R\, y^R \mid x^L\, y + x\, y^R - x^L y^R, x^R y + x\, y^L - x^R\, y^L\}$.

33.   J.H. Conway, 'All Games Bright and Beautiful', *American Mathematics Monthly* 84, pp. 417–34 (1977).

34.   A. Huxley, *Point Counter Point*, Grafton, London (1928), p. 135.

35.   J. Hick, *Arguments for the Existence of God*, Macmillan, London (1970).

36.   Anselm, Proslogion 2.

37.   C. Hartshorne, *A Natural Theology for our Time*, Open Court, La Salle (1967). A fuller discussion and bibliography can be found in J.D. Barrow & F.J. Tipler, *The Anthropic Cosmological Principle*, Oxford University Press (1986), pp. 105–9.

38.   B. Russell, 'Recent work on the principles of mathematics', *International Monthly*, 4 (1901).

39.   G. Cantor, *Grundlagen einer allegemeinen Mannigfaltigkeitslehre*, B.G. Treubner, Leipzig (1883), p. 182; English transl. as *Foundations of the Theory of Manifolds*, transl. U. Parpart, the *Campaigner* (The Theoretical Journal of the National Caucus of Labor Committees), 9, pp. 69–96 (1976). The translation here is from J. Dauben, *Georg Cantor*, Harvard University Press, Mass. (1979), p. 132.

chapter six

Empty Universes

1.   P. Kerr, *The Second Angel*, Orion, London (1998), p. 201.

2.   By 'strong' we mean that the gravitational force gradient can induce particles to move at speeds close to that of light.

3.   Light moves more slowly through a medium than it does through a vacuum. It is possible for objects to travel through a medium at a speed which exceeds the speed of light in that medium. When this occurs then radiation, called Cerenkov radiation, is produced and is routinely observed. It is used by experimenters to detect high-speed particles from space.

4.   We talk of mass and energy together because they are equivalent, related by Einstein's famous formula $E = mc^2$, where E is energy, m is mass and c is the velocity of light in a vacuum.

5.   The ripples are called gravitational waves. They travel at the speed of light and can be viewed as the propagating influence of gravity fields. The long-range effect of rotation, called the dragging of inertial frames, pulls objects around in the same sense as that of the rotation possessed by a nearby source of gravity. Both of these phenomena are absent in Newton's theory of gravity.

6.   Curved space is easy to visualise but curved time sounds strange. In practice it amounts to a change in the rate of flow of time compared to the rate at a place, ideally infinitely far from all masses, where the space is flat. In general, clocks measure time to pass slower in strong gravitational fields than it passes in weak gravitational fields. This is also observed.

7.   An interesting and controversial consequence of this picture is that it implies that the spacetime is the primary concept, rather than space or time separately or added together. The block of spacetime can be sliced up into a stack of curved sheets in an infinite number of different ways, all apparently as good as any other. This corresponds to a choice of time. Events on each slice are simultaneous but different moving observers create different slicings, different

standards of time, and make different observations which they judge to be simultaneous. This block spacetime picture implies that the future is already 'out there'. By contrast, in other sciences, the flow of time is associated with unfolding events, increase in information, entropy or complexity, and there is no suggestion that the future is out there waiting. For an interesting discussion of the theological and philosophical implications of the block spacetime picture, see C.J. Isham and J.C. Polkinghorne, 'The Debate over the Block Universe', in *Quantum Cosmology and the Laws of Nature* (2nd edn), eds R.J. Russell, N. Murphy and C.J. Isham, University of Notre Dame Press (1996), pp. 139–47.

8. All fundamental forces appear to possess 'carrier' (or 'exchange') particles which convey the interaction. The carrier for the electromagnetic interaction between electrically changed particles is the photon which does *not* possess charge and so is not self-interacting. Gravity is carried by the graviton (which is the same as the gravitational waves discussed above) which possesses mass energy and so feels the force of gravity and is self-interacting. You can have a gravitating world that contains only gravitons but not an electromagnetic world that contains only photons.

9. Isaiah 34 v. 11–12.

10. J.D. Barrow, *The Origin of the Universe*, Orion, London (1994).

11. So that as the distribution of mass and energy changes from one slice to another there will be conservation of energy and electric charge and angular momentum.

12. M.J. Rees and M. Begelman, *Gravity's Fatal Attraction*, Scientific American Library, New York (1996), p. 200.

13. C.S. Peirce, *The Collected Papers of Charles Sanders Peirce* (8 vols), ed. C. Hartshorne et al, Harvard University Press, Cambridge, Mass. (1931–50), vol. 4, section 237.

14. E. Mach, *The Science of Mechanics*, first published in 1883, reprinted by Open Court, La Salle (1911).

15. J.D. Barrow, R. Juszkiewicz and D. Sonoda, 'Universal Rotation: How Large Can It Be?', *Mon. Not. Roy. Astr. Soc.*, 213, pp. 917–43 (1985).

16.   The inflationary universe theory, which will be described in the next chapter, leads us to expect that the rotation of the Universe will be very small. Any rotation that existed before inflation (a period when the expansion of the Universe accelerates) occurs will be dramatically reduced during a period of inflation. Moreover, the matter fields expected to produce inflation cannot rotate and so inflation cannot create rotations in the way that it can produce variations in density and in gravitational waves. Indeed, the observation of large-scale rotation in the Universe would be fatal for the inflationary theory, see J.D. Barrow & A. Liddle, 'Is inflation falsifiable?' *General Relativity & Gravitational Journal*, 29, pp. 1501–8 (1997).

17.   Of course, if you are interested in particular questions like how the small deviations from homogeneity and isotropy arose, and why they have the observed patterns, then you do not begin with such an assumption. Instead, you might assume that the irregularities are small (but non-zero) and that the Universe was just homogeneous and isotropic on the average.

18.   A. Friedmann, *Zeitschrift für Physik*, 10, p. 377 (1922) and 21, p. 326 (1924). Translations appear in *Cosmological Constants*, eds J. Bernstein and G. Feinberg, Columbia University Press (1986). R.C. Tolman, *Relativity, Thermodynamics and Cosmology*, Oxford University Press (1934).

19.   The alternative scenario of contraction is ruled out on the grounds that it would have resulted in a 'crunch' of high density already.

20.   Friedmann was a daring balloonist in the cause of science and even held the world altitude record at one time. These flights appear reckless by modern standards, with the balloonists often undergoing calculated periods of unconsciousness in extreme weather conditions. For biographical details of these adventures, see E.A. Tropp, V.Ya. Frenkel and A.D. Chernin, *Alexander A. Friedmann: The Man Who Made the Universe Expand*, transl. A. Dron and M. Burov, Cambridge University Press (1993).

21.   R. Rucker, *The Fourth Dimension*, Houghton Mifflin, Boston (1984), p. 91.

22.   G. Lemaître, 'Evolution of the expanding universe', *Proceedings of the National Academy of Sciences, Washington*, 20, p. 12 (1934).

23.   If a fluid has pressure p and energy density ρc^2, where c is the speed of light, then the condition for its gravitational effect to be attractive (repulsive) is that $\rho c^2 + 3p$ be positive (negative). In a homogeneous and isotropic universe the cosmological constant is equivalent to a 'fluid' with $p = -\rho c^2$ and hence it is gravitationally repulsive.

24.   This description of expanding universes that begin at a past moment of high (infinite?) density was coined pejoratively by Fred Hoyle in a radio broadcast in 1950, to contrast it with the steady-state theory.

25.   W.H. McCrea, *Proc. Roy. Soc. A* 206, p. 562 (1951). Lemaître's early article (ref. 22) on the interpretation of the lambda term as a fluid with pressure and density in the context of general relativity was not known to McCrea.

26.   A. Sandage, *Astrophysical Journal Letters* 152, L 149–154 (1968).

27.   D. Sobel, *Longitude*, Fourth Estate, London (1995).

28.   With a 95% statistical confidence level.

29.   S. Perlmutter et al, 'Measurements of Ω and Λ from 42 high-redshift supernovae', *Astrophysical Journal*, 517, pp. 565–58 (1999) B.P. Schmidt et al, 'The high-Z supernova search: measuring cosmic deceleration and global curvature of the Universe using type Ia supernovae', *Astrophysical Journal*, 507, pp. 46–63 (1998). Updated information about the Supernova Cosmology Project can be obtained from the Project website at http://panisse.lbl.gov/public/papers.

chapter   seven

The Box That Can Never Be Empty

1.   B. Hoffman, *The Strange Story of the Quantum*, Penguin, London (1963), p. 37.

2.   A. Einstein, letter to D. Lipkin, 5 July 1952, quoted in A. Fine, *The Shaky Game*, University of Chicago Press (1986), p. 1.

3. My own version, with many references to others, can be found in J.D. Barrow, *The Universe that Discovered Itself,* Oxford University Press (2000).

4. Quoted by N.C. Panda in *Maya in Physics,* Motilal Bonarsidass Publishers, Delhi (1991), p. 73.

5. A. Einstein, letter to Max Born, 4 June 1919, quoted by Max Born in *The Born–Einstein Letters,* Walker & Co., New York (1971), p. 11.

6. R. Feynman, *The Character of Physical Law,* MIT Press, Cambridge, Mass. (1967), p. 129.

7. W. Heisenberg, *Physics and Beyond: Encounters and Conversations,* Harper and Row, New York (1971), p. 210.

8. H.A. Kramers, quoted in L. Ponomarev, *The Quantum Dice,* IOP, Bristol (1993), p. 80.

9. Black bodies are perfect absorbers and emitters of light.

10. Zero degrees Centigrade equals 273·15 degrees Kelvin.

11. Its numerical value is measured to be $h = 6.626 \times 10^{-34}$ Joule-seconds.

12. It was predicted that the spectrum should have a Planckian shape over most of the wavelength range but there was great interest in how accurately it would follow the Planck curve in certain wavelength ranges. This interest arose because, if the history of the Universe had undergone violent episodes associated with the formation of galaxies, other sources of radiation with higher temperatures could have been added to the primeval radiation left over from the Big Bang. This can distort the spectrum slightly from the Planckian form. The observations showed no such distortions of the pure Planck spectrum to very high precision. This tells us important things about the history of the Universe.

13. J.C. Maxwell, *Treatise on Electricity and Magnetism,* Dover, NY (1965).

14. The zero-point energy idea was first introduced by Planck in 1911 in an attempt to understand how matter and radiation interact to create the black-body Planck spectrum. Planck first proposed that whilst the emission of radiation occurs in discrete quantum packets the absorption of radiation is continuously possible over all values.

This hypothesis (which Planck abandoned three years later) led to the conclusion that the system would have energy hf/2 even at absolute zero of temperature. In 1913 Einstein and Otto Stern showed that the correct classical (non-quantum) limit for the energy is only obtained from the Planck black-body distribution if the zero-point energy is included (*Annalen der Physik*, 40, pp. 551–60 [1913]). For some further discussion see D.W. Sciama, in *The Philosophy of Vacuum*, Oxford University Press (1991), pp. 137–58.

15.   Casimir, H.B.G., 'On the attraction between two perfectly conducting plates', *Koninkl. Ned. Akad. Wetenschap. Proc.*, 51, pp. 793–5 (1948); Casimir's first study of these effects dealt with the more specific situation of the attractive force between two polarisable atoms. An attractive force arises and led Casimir to replace the atoms by the simpler situation of parallel plates. H.B.G. Casimir and D. Polder, 'The Influence of Retardation on the London-van der Waals Forces', *Phys. Rev.* 73, pp. 360–72 (1948); G. Plumien, B. Muller and W. Greiner, 'The Casimir Effect', *Phys. Rep.*, 134, pp. 87–193 (1986). A complete calculation of the effect requires several important details to be taken into account, for example the fact that the plates cannot be regarded as perfect conductors down on the scale of single atoms and smaller. The most complete book on the subject is P.W. Milonni, *The Quantum Vacuum: an introduction to quantum electrodynamics*, Academic, San Diego (1994). A simple account can be found in T. Boyer, 'The classical vacuum', *Scientific American* (Aug. 1985).

16.   The formula gives 0.02 Newtons per square metre. These numbers are taken from the experimental investigation of Sparnaay.

17.   J. Ambjorn and S. Wolfram, *Ann. Phys.*, 147, p. 1 (1983); G. Barton, 'Quantum electrodynamics of spinless particles between conducting plates', *Proc. Roy. Soc. A*, 320, pp. 251–75 (1970).

18.   M.J. Sparnaay, 'Measurement of the attractive forces between flat plates', *Physica*, 24, p. 751 (1958).

19.   S.K. Lamoreaux, 'Demonstration of the Casimir force in the 0.6 to 6µM range', *Phys. Rev. Lett.* 78, pp. 5–8 (1997) and 81, pp. 5475–6 (1998).

20. Careful account must be taken of the fact that the experiment is not being performed at absolute zero of temperature and that the plates (made of coated quartz) are not perfect conductors as assumed in the simple calculation we have described.

21. C.I. Sukenik, M.G. Boshier, D. Cho, V. Sandoghdar and E. Hinds, 'Measurement of the Casimir–Polder force', *Phys. Rev. Lett.*, 70, pp. 560–3 (1993).

22. H.E. Puthoff, 'Gravity as a zero-point fluctuation force', *Phys. Rev. A*, 39, pp. 2333–42 (1989); R.L. 'Forward, Extracting electrical energy from the vacuum by cohesion of charged foliated conductors', *Phys. Rev. B*, 30, pp. 1700–2 (1984); D.C. Cole & H.E. Puthoff, 'Extracting energy and heat from the vacuum', *Phys. Rev. E*, 48, pp. 1562–5 (1993); I.Y. Sokolov, 'The Casimir Effect as a possible source of cosmic energy', *Phys. Let. A*, 223, pp. 163–6 (1996); P. Yam, 'Exploiting zero-point energy', *Scientific American*, 277, pp. 82–5 (Dec. 1997).

23. J. Schwinger, 'Casimir light: field pressure', *Proc. Nat. Acad. Sci. USA*, 91, pp. 6473–5 (1994); C. Eberlein, 'Sonoluminescence as quantum vacuum radiation', *Phys. Rev. Lett.*, 76, pp. 3842–5 (1996).

24. K.A. Milton and Y.J. Ng, 'Observability of the bulk Casimir effect: can the dynamical Casimir effect be relevant to sonoluminescence?', *Phys. Rev. E*, 57, pp. 5504–10 (1998); V.V. Nesterenko and I.G. Pirozhenko, 'Is the Casimir effect relevant to sonoluminescence?', *Sov. Physics JETP Lett.*, 67, pp. 420–4 (1998).

25. J. Masefield, *Salt-water Ballads*, 'Sea Fever' (1902).

26. P.C. Causeé, *L'Album du Marin*, Charpentier, Nantes (1836).

27. S.L. Boersma, 'A maritime analogy of the Casimir effect', *American J. Physics*, 64, p. 539 (1996). The author says that his attention was drawn to this problem in Causeé's book by Hazelhoff Roelfzema of the Amsterdam Shipping Museum.

28. The force of attraction calculated by Boersma is equal to $F = 2\pi^2 m\eta h A^2/(QT^2)$ where m is the mass of each ship (the two ships are assumed to be of equal mass), A is the angle in radians of their rolling in the swell, h is the metacentric height of the ship, T is the

period of the oscillations of the ships, Q the quality factor of the oscillation, and η is the efficiency of the energy losses through friction. Substituting m = 700 tons, h = 1.5 metres, A = 8 degrees (= 0.14 radians), T = 8 seconds, Q = 2.5, η = 0.8 we find F = 2000 Newtons.

29.   In J. Weintraub, *Peel Me a Grape* (1975), p. 47.

30.   W. Lamb and R.C. Retherford, *Phys. Rev.*, 72, p. 241 (1947). The theoretical interpretation was supplied by T.A. Welton, *Phys. Rev.*, 74, p. 1157 (1948).

31.   P. Kerr, *The Second Angel*, Orion, London (1998), p. 316.

32.   To see what else is needed to understand the world, see J.D. Barrow, *Theories of Everything*, Vintage, London (1988).

33.   Because the quantum wavelength of a particle is inversely proportional to its mass.

34.   F. Close and C. Sutton, *The Particle Connection*, Oxford University Press (1987).

35.   Despite its intrinsic weakness gravity wins out over electromagnetism in controlling the behaviour of matter in large aggregates because electric charges come in two varieties, positive and negative, and it is hard to assemble a large amount of matter that has a non-zero charge. Gravity acts on mass, and mass, by contrast, comes only in a positive variety and so its effect is cumulative when large aggregates of material are assembled.

36.   Lao-tzu, *Tao Te Ching*, chap. 11.

37.   This number, first defined by Arnold Sommerfeld in 1911, is called the fine structure constant and is given by $2\pi e^2/hc$. For further details of these developments see Chapter 4 of J.D. Barrow and F.J. Tipler, *The Anthropic Cosmological Principle*, Oxford University Press (1986).

38.   C. Pickover, *Computers and the Imagination*, St Martin's Press, NY (1991), p. 270.

39.   The positron is the antiparticle of the electron. It has the same mass but opposite sign of its electric charge. When an electron encounters a positron they will annihilate to produce two photons of light. The electric charges cancel out to zero.

40. The term 'black hole' was invented by the American physicist John A. Wheeler in 1968 in an article entitled 'Our Universe: the known and the unknown', *American Scholar*, 37, p. 248 (1968), reprinted in *American Scientist*, 56 p. 1 (1968). Eleven years earlier he had coined the term 'wormhole'.

41. The radius of a black hole, R, is proportional to its mass, M, therefore its density, proportional to M/R^3, varies as $1/M^2$. Hence, the more massive the black hole, the lower its density.

42. The 'killer' feature of a black hole is the strong variation in gravitational pull that it exerts over an object of a finite (the 'tidal force'). For a point particle of zero size no such variation exists and it would feel nothing as it fell freely under gravity into a black hole, big or small. Objects of finite size get stretched out because the part of them closest to the hole gets pulled in more strongly than the part furthest away. The density of the black hole (see the previous footnote) is a good measure of the strength of this tidal force. It becomes more significant for small black holes. Black holes smaller than about one hundred million times the mass of our Sun are able to tear stars apart when they fall through the horizon. Bigger black holes allow whole stars to fall through the horizon without breaking them up.

43. J.P. Luminet, *Black Holes*, Cambridge University Press (1992).

44. M. Begelman and M.J. Rees, *Gravity's Fatal Attraction*, W.H. Freeman, San Francisco (1996).

45. S.W. Hawking, 'The Quantum Mechanics of Black Holes', *Scientific American*, January (1977).

46. The black hole does not gain mass as a result of capturing one of the pair of particles. The black hole loses mass as a result of this process after one takes into account the change in the potential energy of the particle-antiparticle pair. The pairs will preferentially appear at a separation that gives them zero total energy.

47. The temperature of the black hole is inversely proportional to its mass. The time required to radiate away all of its mass is proportional to the cube of its mass.

48.   In Einstein's theory of gravitation the local effects of gravity should be indistinguishable from those experienced by undergoing accelerated motion at a suitable rate. Thus, for very short intervals of time, we should not be able to distinguish being in a small lift that is freely falling under gravity from one that is being accelerated downwards. If we apply this to the situation of black holes in the vacuum, we should not be able to distinguish the situation that exists in the gravitational field of the black hole very close to the event horizon from that experienced by an observer moving with an acceleration equal to the acceleration due to gravity at the horizon. In fact, as first shown by Bill Unruh and Paul Davies, this is exactly what is predicted. If an observer is accelerated through the quantum vacuum at a uniform rate, A, then that observer should detect thermal radiation with a temperature given by $T = hA/4\pi ck$, where c is the speed of light, k is Boltzmann's constant and h is Planck's constant.

chapter   eight

How Many Vacuums Are There?

I.   Quoted in the *Observer* newspaper, 4 July 1999.

2.   Contrary to the situation that exists in many other subjects, research journals are now largely superfluous in subjects like physics and astronomy. All research articles are 'published' electronically and are revised in the light of the comments, requests for credit, or corrections, that come in by email to the authors.

3.   Note that Newton's famous laws of motion did not satisfy this dictum of Einstein's. The first law, which was that 'all bodies acted upon by no forces remain at rest or move at constant speed', is not one that would be found true by all observers. Newton specified that it would be seen only by observers who were not accelerating or rotating with respect to absolute space. These are known as 'inertial observers'. For example, if an astronaut in a rotating spaceship were to look out of the window he would see a neighbouring satellite

accelerating past his window even if it were acted upon by no forces. The astronaut in his rotating spaceship is not an inertial observer.

4. Quoted in *Observer*, 12 December 1999, p. 30.

5. J.D. Barrow, *Theories of Everything*, Vintage, London (1991); B. Greene, *The Elegant Universe*, Vintage, London (2000).

6. A. Linde, 'The Self-Reproducing Inflationary Universe', *Scientific American*, no. 5, vol. 32 (1994).

7. A. Guth, *The Inflationary Universe*, Vintage, London (1998).

8. Fractal surfaces do not. They can have structure on all scales of magnification.

9. W. Allen, *Getting Even*, Random House, NY (1971), p. 33.

10. This hierarchy of clustering of clusters does not carry on indefinitely. The clustering of galaxy clusters into so-called 'superclusters' seems to be the end of the line.

11. A term coined by the sociologist Robert Merton to describe the way in which people who are awarded honours and prizes then seem to be awarded even more honours and prizes. It is taken from Christ's words in the Gospel of St Matthew chap. 13 v. 12: 'For whosoever hath, to him shall be given, and he shall have more abundance: but whosoever hath not, from him shall be taken away even that he hath.'

12. When it was about ten million years old.

13. P. de Bernadis et al, 'A flat universe from high-resolution maps of the cosmic microwave background radiation', *Nature*, 404, pp. 955–9 (2000). See also http://www.physics.ucsb.edu/~boomerang/ for further pictures and information about the experiment and the significance of its results.

14. The *Independent*, quoted in third Leader article in Review section, p. 3, 13 November 1999.

15. Based on data presented by the Boomerang Collaboration on their website http://www.physics.ucsb.edu/~boomerang/.

16. J.D. Barrow, 'Dimensionality', *Proc. Roy. Soc. A.*, 310, p. 337 (1983); J.D. Barrow & F.J. Tipler, *The Anthropic Cosmological Principle*, Oxford University Press (1986), chap. 6; M. Tegmark, 'Is "the theory of

everything" merely the ultimate ensemble theory?', *Annals of Physics (NY)*, 270, pp. 1–51 (1998).

17.   There was once some interest in science-fiction stories about bio-chemists based upon silicon chemistry. These do not look promising (as explained in Barrow and Tipler, op. cit.), but, ironically, it is silicon *physics* that looks the most likely route to a form of artificial life brought into being by means of the catalytic help of (human) carbon-based life.

18.   During inflation the pressure contributed by the scalar matter field responsible is negative and so a change in energy of the expanding material actually provides energy for work.

19.   A few years ago Sidney Coleman proposed a partial solution of this sort. He suggested that if the topology of the Universe was sufficiently complicated, with many holes, handles and tubes ('wormholes'), then the presence of any lambda term would tend to create an opposing stress that cancelled out the lambda term to very high precision. The most probable value of lambda that would be measured when the Universe expanded and became very large would be zero with great accuracy. Unfortunately, the appealing idea did not survive further scrutiny and there are no similar general arguments which have so far provided us with an understanding of lambda's tiny value.

20.   I Corinthians chap. 15 v. 51–2.

21.   The later editions of inflation, like chaotic inflation, could make do with a single vacuum.

22.   P. Hut and M.J. Rees, 'How stable is our vacuum?', *Nature*, 302 (1983), pp. 508–9; M.S. Turner and F. Wilczek, 'Is our vacuum metastable?', *Nature* 298 (1982), pp. 633–4. For a wider review of possible 'sudden' ends to the world see J. Leslie, *The End of the World*.

23.   Recently, there seems to have been some public worry in the United States that a planned sequence of high-energy particle collisions at a national laboratory might induce just such a catastrophe.

24.   N. Eldridge and S.J. Gould, 'Punctuated equilibria: an alternative to phyletic gradualism', in T.J.M. Schoof (ed.), *Models in Paleobiology*, W.H. Freeman, San Francisco (1972), pp. 82–115.

25. From P. Bak, *How Nature Works*, Oxford University Press (1997), p. 39.

26. A.S.J. Tessimond, *Cats*, p. 20 (1934).

27. T. Kibble, 'Topology of Cosmic Domains and Strings', *Journal of Physics A*, 9, pp. 1387–97 (1972).

28. These should not be confused with superstrings or superstring theories. Superstring theories may permit the existence of cosmic strings but not necessarily.

29. From P. Bak, op. cit.

30. This gravitational lensing phenomenon, predicted by Einstein, is now commonly observed but is believed to be created by objects other than cosmic strings in the cases where it is seen. In our own galaxy and nearby in the Large Magellanic Cloud (a neighbouring very small galaxy) it is created by non-luminous objects that have masses similar to those of stars.

chapter nine

The Beginning and the End of the Vacuum

1. M. Proust, *Le Côté de Guermantes* (1921), transl. as *Guermantes' Way*, by C.K. Scott-Moncrieff (1925), vol. 2, p. 147.

2. G.K. Chesterton, *The Napoleon of Notting Hill*, first published in 1902, Penguin, London (1946), p. 9.

3. E.O. James, *Creation and Cosmology*, E.J. Brill, Leiden (1969); C. Long, *Alpha: the Myths of Creation*, G. Braziller, New York (1963); C. Blacker and M. Loewe (eds), *Ancient Cosmologies*, Allen & Unwin, London (1975); M. Leach, *The Beginning: Creation Myths around the World*, Funk and Wagnalls, New York (1956).

4. M. Eliade, *The Myth of the Eternal Return*, Pantheon, New York (1954); see also J.D. Barrow & F.J. Tipler, *The Anthropic Cosmological Principle*, Oxford University Press (1986).

5. T. Joseph, 'Unified Field Theory', *New York Times*, 6 April 1978.

6. An interesting collection of articles is to be found in R. Russell, N. Murphy & C. Isham, *Quantum Cosmology and the Laws of Nature*, 2nd edn, University of Notre Dame Press, Notre Dame (1996).

7.   A. Ehrhardt, 'Creatio ex Nihilo', *Studia Theologica* (Lund), 4, p. 24 (1951), and *The Beginning: A Study in the Greek Philosophical Approach to the Concept of Creation from Anaximander to St. John*, including a memoir by J. Heywood Thomas, Manchester University Press (1968); D. O'Connor and F. Oakley (eds), *Creation: the impact of an idea*, Scribners, New York (1969).

8.   With St Augustine, this idea was made more sophisticated by including the idea that time must have come into being along with the world, thus avoiding questions about what existed 'before' the world was.

9.   S. Jaki, *Science and Creation*, Scottish Academic Press, Edinburgh (1974), and *The Road of Science and the Ways to God*, University of Chicago Press (1978).

10.   G.F. Moore, *Judaism in the First Centuries of the Christian Era I*, Cambridge, Mass. (1966), p. 381.

11.   Wisdom chap. 11 v. 17.

12.   7 v. 28.

13.   G. May, *Creatio ex Nihilo*, transl. A.S. Worrall, T & T Clark, Edinburgh (1994), p. 8.

14.   H.A. Wolfson, 'Negative Attributes in the Church fathers and the Gnostic Basilides', *Harvard Theol. Review*, 50, pp. 145–56 (1957), J. Whittaker, 'Basilides and the Ineffability of God', *Harvard Theol. Review*, 62, pp. 367–71 (1969).

15.   A beautiful expression of this is found in the Tripartite Tractate of Valentius in the Jung Codex, cited by May, op.cit., p. 75, translation by H.W. Attridge and E. Pagels: 'No one else has been with him from the beginning; nor is there a place in which he is, or from which he has come forth, or into which he will go; nor is there a primordial form, which he uses as a model as he works; nor is there any difficulty which accompanies him in what he does; nor is there any material which is at his disposal, from which he creates what he creates; nor any substance within him from which he begets what he begets; nor a co-worker with him, working with him on the things at which he works. To say anything of this sort is ignorant.'

16. Four centuries later the argument will still be used by John Philoponus to counter the same suggestions. He defines the creativity of the artist as an activity which rearranges existing elements in a new way and the creativity of the natural world as the bringing of living beings out of non-living matter. Divine creation is superior to both because it can create the material out of nothing.

17. Nevertheless, other aspects of his view of the world were different to those that would be adopted by the central Christian tradition. Basilides appears to have been a Deist in believing that God played no further role in the unfolding of the Universe after laying down the starting conditions. God's creative activity was limited to a single act.

18. F.M. Cornford, *Microcosmographia Academica*, Cambridge University Press (1908), p. 28.

19. See N. Rescher, *The Riddle of Existence*, University Press of America, Lanham (1984), p. 2.

20. J.D. Barrow, *Impossibility*, Oxford University Press (1998).

21. N. Malcolm, *Ludwig Wittgenstein: A Memoir*, Oxford University Press (1958), p. 20.

22. M. Heidegger, *Introduction to Metaphysics*, Yale University Press, New Haven (1959); L. Wittgenstein, *Tractatus Logico-Philosophicus*, London (1922), section 6.44.

23. H. Bergson, *Creative Evolution*, trans. A. Mitchell, Modern Library, NY (1941), p. 299. This type of set-theoretic basis for conjuring something out of nothing is also hinted at in the discussion to be found in the final chapter of *Gravitation* by C. Misner, K. Thorne & J.A. Wheeler, W.H. Freeman, San Francisco (1972); and in P. Atkins, *The Creation*, W.H. Freeman, San Francisco (1981).

24. For example, the three interior angles of a triangle sum to 180 degrees in a Euclidean geometry but not in a non-Euclidean geometry.

25. A. Hodges, *The Enigma of Intelligence*, Unwin, London (1985), p. 154.

26. J.D. Barrow, *Impossibility*, Oxford University Press (1998).

27. R. Penrose, *The Emperor's New Mind*, Oxford University Press (1989).

28.   The feature that there exist statements of arithmetic whose truth or falsity cannot be established using the rules and axioms of arithmetic; see J.D. Barrow, *Impossibility*, Oxford University Press (1998), for a fuller discussion.

29.   N. Rescher, *The Riddle of Existence*, University Press of America, Lanham (1984), p. 3.

30.   T. Joseph, 'Unified Field Theory', *New York Times*, 6 April 1978.

31.   Einstein thought that infinities and singularities were unacceptable in physical theories. His assistant at Princeton, Peter Bergmann, writes: 'It seems that Einstein always was of the opinion that singularities in classical field theory are intolerable. They are intolerable from the point of view of classical field theory because a singular region represents a breakdown of the postulated laws of nature. I think that one can turn this argument around and say that a theory that involves singularities and involves them unavoidably, moreover, carries within itself the seeds of its own destruction.' Contribution in H. Woolf, *Some Strangeness in the Proportion*, Addison Wesley, MA (1980), p. 156.

32.   For a recent overview of the mathematical ideas see the first chapter of S.W. Hawking & R. Penrose, *The Nature of Space and Time*, Princeton University Press (1996). For a descriptive account see J.D. Barrow & J. Silk, *The Left Hand of Creation* (2nd edn.), Penguin, London, and Oxford University Press, New York (1993).

33.   The recently discovered microwave background radiation was sufficient to meet this requirement.

34.   It is known that incomplete histories occur which are not accompanied by infinities of physical quantities, like density or temperature. However, whilst it is suspected that these examples are in some way atypical of solutions to Einstein's equations, this has not been proven in general. The original theorem of Penrose was proved for the situation of a collapsing cloud of matter (like the expanding universe in reverse). Subsequently, Hawking and Penrose proved a version of the theorem which applies specifically to cosmologies.

For a detailed survey, see S.W. Hawking & G.F.R. Ellis, *The Large Scale Structure of Space-time*, Cambridge University Press (1973).

35. J. Earman, *Bangs, Crunches, Whimpers, and Shrieks: singularities and acausalities in relativistic spacetimes*, Oxford University Press (1995).

36. If one interprets the lambda force as a vacuum energy in Einstein's equations then it behaves like a form of matter that exhibits gravitational repulsion because its pressure p and density ρ satisfy the relationship $p = -\rho c^2$. Gravitational repulsion arises whenever matter satisfies the weaker condition $3p < -\rho c^2$. The singularity theorems assume that $3p > -\rho c^2$.

37. M. Eliade, *The Myth of the Eternal Return*; see also J.D. Barrow & F.J. Tipler, *The Anthropic Cosmological Principle*, Oxford University Press (1986).

38. It is very likely that these two singularities would be very different in structure. Irregularities tend to grow during the evolution of the Universe in its expanding phase. These irregularities will be amplified even further during the contraction phase and the final singularity should be extremely irregular.

39. The big assumption here is that nothing counter to the second law of thermodynamics occurs at the moments when the Universe bounces (or indeed, that any such 'law' is applicable).

40. This was first pointed out by R.C. Tolman in two articles, 'On the problem of the entropy of the universe as a whole', *Physical Review*, 37, pp. 1639–1771 (1931), and 'On the theoretical requirements for a periodic behaviour of the universe', *Physical Review*, 38, p. 1758 (1931). Recently, a detailed reanalysis was given by J.D. Barrow and M. Dąbrowski, 'Oscillating Universes', *Mon. Not. Roy. Astron. Soc.*, 275, pp. 850–62 (1995).

41. See, for example, E.R. Harrison, *Cosmology: the science of the universe*, Cambridge University Press (1981), pp. 299–300.

42. A. Swinburne, 'The Garden of Proserpine', *Collected Poetical Works*, p. 83.

43. F. Dyson, 'Life in an open universe', *Reviews of Modern Physics*, 51, p. 447 (1979).

44.   J.D. Barrow & F.J. Tipler, *The Anthropic Cosmological Principle*, Oxford University Press (1986), chap. 10.

45.   The absolute minimum amount of energy required to process a given amount of information is determined by the second law of thermodynamics. If ΔI is the number of bits of information processed, the second law requires $\Delta I \leq \Delta E/kT\ln 2 = \Delta E/T(\text{ergs/K})(1.05 \times 10^{16})$, where T is the temperature in degrees Kelvin, k is Boltzmann's constant and ΔE is the amount of free energy expended. If the temperature operates at a temperature above absolute zero ($T > 0$, as required by the third law of thermodynamics), there is a minimum amount of energy that must be expended to process a single bit of information. This inequality is due to Brillouin.

46.   See S.R.L. Clark, *How to Live Forever*, Routledge (1995).

47.   See *The Anthropic Cosmological Principle*, op.cit., p. 668.

48.   The current observations are indicating that this is not the case in our Universe. It appears to be destined to keep expanding for ever, locally, and if the eternal inflation scenario is true it will continue expanding globally as well. Recently, João Magueijo, Rachel Bean and I ('Can the Universe escape eternal inflation?', *Mon. Not. Roy. Astron. Soc.*, 316, L41–44 [2000]) have found a way for the Universe to escape from accelerating expansion. If it contains a scalar field, which is falling in a potential energy landscape which descends steeply but has a small U-shaped crevice on it, with a local minimum, then the scalar field can pass through along this valley and produce a short period of inflation. It carries on up the slope and then continues to fall down the slope again. When this happens the expansion stops accelerating and reverts to the usual decelerated expansion that it experiences for most of its history. Potential landscapes with this shape have been identified in string theories at high energy. They were suggested for cosmological applications by A. Albrecht and C. Skordis, *Phys. Rev. Lett.*, 84, pp. 2076–9 (2000), but they envisaged that they would lead to a state of never-ending inflation.

49. This must lie at least about thirty billion years in the future. It should be noted that it is possible for us to encounter a singularity in the future without this lambda energy decay, even if the expansion appears to be going to carry on for ever. There could be a gravitational shock-wave travelling towards us at the speed of light that hits us without warning.

Index